SpringerBriefs in Geography

SpringerBriefs in Geography presents concise summaries of cutting-edge research and practical applications across the fields of physical, environmental and human geography. It publishes compact refereed monographs under the editorial super-vision of an international advisory board with the aim to publish 8 to 12 weeks after acceptance. Volumes are compact, 50 to 125 pages, with a clear focus. The series covers a range of content from professional to academic such as: timely reports of state-of-the art analytical techniques, bridges between new research results, snapshots of hot and/or emerging topics, elaborated thesis, literature reviews, and in-depth case studies.

The scope of the series spans the entire field of geography, with a view to significantly advance research. The character of the series is international and multidisciplinary and will include research areas such as: GIS/cartography, remote sensing, geographical education, geospatial analysis, techniques and modeling, landscape/regional and urban planning, economic geography, housing and the built environment, and quantitative geography. Volumes in this series may analyze past, present and/or future trends, as well as their determinants and consequences. Both solicited and unsolicited manuscripts are considered for publication in this series.

SpringerBriefs in Geography will be of interest to a wide range of individuals with interests in physical, environmental and human geography as well as for researchers from allied disciplines.

More information about this series at http://www.springer.com/series/10050

Timothy Tambassi

The Philosophy of Geo-Ontologies

Applied Ontology of Geography

Second Edition

 Springer

Timothy Tambassi
Department of Science of Cultural Heritage
University of Salerno
Fisciano, Italy

ISSN 2211-4165 ISSN 2211-4173 (electronic)
SpringerBriefs in Geography
ISBN 978-3-030-78144-6 ISBN 978-3-030-78145-3 (eBook)
https://doi.org/10.1007/978-3-030-78145-3

This Springer imprint is published by the registered company Springer Nature Switzerland AG
The registered company address is: Gewerbestrasse 11, 6330 Cham, Switzerland

Introduction

Placed at the intersection among philosophy, geography, and computer science, the domain of investigation of applied ontology of geography ranges

- from making explicit assumptions and commitments of geography as a discipline;
- to the theoretical and technical needs of geographical/IT tools, such as GIS and geo-ontologies.

Such a domain of investigation represents the central topic of discussion of this book, which intends:

- to provide an overview of the mutual interactions among the disciplines encompassed in the domain;
- to discuss notions such as spatial representation, boundaries, and geographical entities that constitute the main focus of the (philosophical) ontology of geography;
- to propose a geographical classification of geo-ontologies in response to their increasing diffusion within the contemporary debate, as well as to show what ontological categories best systematize their contents.

The book is divided into three parts. In the first one, *Among Computer Science, Philosophy, and Geography: An Ontological Investigation*, Chap. 1 explores the domain of investigation of applied ontology of geography, showing how the disciplines involved are connected to one another and outlining some possible strategies to provide a way out of the disconnections. Chapter 2 analyzes the kinds of ontologies behind the applied ontology of geography. First, it concerns IT/computer science, within which ontologies are generally conceived as explicit specifications of shared conceptualizations. Second, philosophical ontology is presented as the research area that pinpoints the totality of entities which make up the world on different levels of focus and granularity and whose different parts and aspects are studied by folk and scientific disciplines. Third, the ontology of geography is distinguished by its focus on establishing what geographical entities exist, developing a theory of spatial representation, and explaining how the geographic descriptions of reality emerging from common sense can be combined with those derived from academic geography.

The second part, *Systematizing the Geographical World*, addresses the main philosophical issues of ontology of geography, as well as the ontological assumptions and commitments of geography, namely spatial representation, boundaries, and geographical entities. More precisely, Chap. 3 undertakes to offer an introduction to the theoretical tools needed for advancing a formal theory of spatial representation, tools which include mereology, topology, and the geo-ontological distinction between classical and non-classical geographies. Chapter 4 focuses on geographical boundaries, with the aim of analyzing how the notion of boundary has been conceived by contemporary geo-ontologists, what kinds of geographical boundaries have been identified and categorized, as well as the influence of cultural diversities and human beliefs upon such categorizations. Finally, Chap. 5 provides a sketch of possible approaches, response attempts, and issues arising from the question: "What is a geographical entity?". The answer to this question will be multi-faceted and will fight the prevailing philosophical trend of simplifying the endless diversity and variation among different geographical perspectives.

While by discussing notions of spatial representation, boundaries, and geographical entities, the second part of the book is mainly speculative, the third part, *The Philosophy of Geo-Ontologies*, analyzes geo-ontologies as an IT/computer application of the theoretical investigation presented above. Chapter 6 explores the emergence of geo-ontologies from the spatial turn and outlines a taxonomy of geo-ontologies grounded on the distinction between spatial, physical, and human geography. The idea behind the taxonomy is to relate geo-ontologies to the geographical debate which, in turn, could improve the conceptualizations of such ontologies. Chapter 7 combines assumptions and requirements coming from IT/computer ontologies, geography, and philosophical ontology, in order to show what categories might complete the current domain of geo-ontologies. The issue is approached by thinking of such a domain as a whole composed of two different levels of categorization. The first level concerns the IT components shared by different ontologies. The second level deals with contents for which philosophical and geographical analysis can include categories that do not appear at the first level. Because the book's concerns are interrelated in numerous and complex ways, I have considered it appropriate to remind the reader in various places of key features of applied ontology of geography, deeming a certain amount of repetition preferable to relying on cross-references too frequently.

The second edition of the book differs from the first one as it offers a broader analysis of the (philosophical) ontology of geography (see, in particular, Chaps. 1 and 3–5): an analysis that is no more limited to the theoretical need of geo-ontologies. The introduction of the subtitle *Applied Ontology of Geography* is meant to remark this change. Some chapters of the book are based upon papers of mine that have been published during the last few years. In particular, Chaps. 1, 3–5, and 7 draw upon the following papers respectively, and I am grateful to the publishers concerned for permission to use the material in this way: Applied Ontology of Geography. Mapping the Interdisciplinary (Un-) Connections. In Tambassi T., Tanca M. (eds.) The Philosophy of Geography. Springer, Cham, 2021; On the Distinction between Classical and Nonclassical Geographies: Some Critical Remarks. In Tambassi T. (ed.)

The Philosophy of GIS (pp. 125–134). Springer, Cham (2019); From Geographical Lines to Cultural Boundaries. Mapping the Ontological Debate. Rivista di Estetica 67:150–64 (2018); What a Geographical Entity Could Be. In Tambassi T. (ed.) The Philosophy of GIS (pp. 177–205). Springer, Cham (2019); What kind of ontological categories for geo-ontologies? Acta Analytica 34(2):135–44 (2019). Moreover, the taxonomy of Chap. 6 is the result of a collaboration with Diego Magro and was originally published in Ontologie informatiche della geografia. Una sistematizzazione del dibattito contemporaneo. Rivista di estetica 58:191–205 (2015).

Acknowledgements. This book is supported by a research fellowship at the Department of Science of Cultural Heritage of the University of Salerno, Grant AIM1873471-2 CUPD44I18000380006. I am extremely grateful to Robert Doe of Springer for all his encouragement and assistance. My debts to other people are too numerous to list. However, I should like to record my particular gratitude to Raffaella Afferni, Margherita Azzari, Ferruccio Andolfi, Armando Bisogno, Beatrice Centi, Sorin Cheval, Fabio Ciotti, Renato de Filippis, Matthew R. X. Dentith, Mihnea Dobre, Giulio d'Onofrio, Anas Fahad Khan, Maurizio Lana, Giulia Lasagni, Giacomo Lenzi, Diego Magro, Cristina Meini, Davide Monaco, Iulia Nitescu, Roberto Poli, Anke Strüver, Marcello Tanca, Italo Testa, Cristina Travanini, Achille Varzi, and Paola Zamperlin for providing comments and feedback, and for their invaluable support. I dedicate this book to Giulia and to our gray paradise, just the way we like it.

Fisciano, Italy   Timothy Tambassi
May 2021

Contents

**Part I   Among Computer Science, Philosophy, and Geography:
An Ontological Investigation**

**1   From the Philosophies of Geographies to the Applied Ontology
of Geography** 3
   1.1   On the Connections Between Philosophy and Geography 3
   1.2   Two Triangles 5
   1.3   Ontologies of Geography 7
   1.4   Digital Geographies 8
   1.5   Big-O Ontology Versus Small-o Ontologies 9
   1.6   The Geo-Ontological Circle 10
   1.7   (Small-o) Ontological Interdisciplinary Connections 11
   1.8   Five Features of (Big-O) Ontological Knowledge 12
   1.9   (Big-G) Geographical Disconnections 13
   1.10   Representations and Practices 14
   1.11   From Applied o/Ontology of Geography to Applied
       o/Ontology of Geography 15
   References 17

2   The Ontological Background 21
   2.1   IT/Computer Ontologies and Semantic Web 21
   2.2   Some Definitions of IT/Computer Ontology 23
   2.3   Clarification of Terms 24
   2.4   From the Ontological Turn to Taxonomies of Philosophical
       Ontologies 25
   2.5   Ontology, Metaphysics, and Scientific Disciplines 26
   2.6   Ontology of Geography 28
   2.7   Common-Sense Geography 29
   2.8   Primary and Secondary Theories 30
   2.9   Mark and Smith's Experiments 31
   2.10   Analysis of Results 32
   References 34

Part II Systematizing the Geographical World

3 Spatial Representation 39
   3.1   Ontology of Geography and Spatial Representation 39
   3.2   Tools for Spatial Representation 40
   3.3   Classical and Non-Classical Geographies 41
   3.4   Issues from Cartographic Representation 42
   3.5   The Capital of Singapore 43
   3.6   Looking for No Man's Land 43
   3.7   Sailing to Thule 44
   3.8   Poland into Exile 45
   3.9   Conclusion 46
   References 46

4 Boundaries 49
   4.1   The Ontology of Geographical Boundaries 50
   4.2   Bona Fide and Fiat Boundaries 51
   4.3   Legal Fiat Boundaries and Normativity 53
   4.4   Physical and Institutional Boundaries 54
   4.5   Boundaries, Cultural Diversities, Human Beliefs 55
   4.6   Cultural Boundaries? 57
   4.7   Three Levels of Cultural Dependence 58
   4.8   Conclusion 59
   References 60

5 Geographical Entities 63
   5.1   A Chaotic List that Cries Out for Explanation 63
   5.2   Avoiding Univocal and Incomplete Accounts 64
   5.3   Laundry Lists 66
   5.4   Attempts of Definition 67
   5.5   On Being Portrayed on Maps 68
   5.6   Maps, Granularity of Interest, and Multiple Levels of Details 69
   5.7   On What and Where 70
   5.8   Drawing the Contour 71
   5.9   Cultural Entities 72
   5.10 GIS, Knowledge Engineering, and Geographic Objects 74
   5.11 Rivers, *Fleuves*, and *Revières* 75
   5.12 Danube, *Donau,* and *Дунав* 76
   5.13 Vagueness 77
   5.14 (Geographical) Kinds and Properties 78
   5.15 Relations, Fields, and Time 79
   5.16 Boundaries 80
   5.17 On Non-Existent and Abstract Geographical Entities 81
   5.18 Historical Entities 82
   5.19 Complex Geographical Entities 83
   5.20 Hierarchical Structures 84

5.21   Three Thin Red Lines   85
5.22   From Multiple (Ways of Doing) Geographies to Multiple
        (Kinds of) Geographical Entities   87
References   88

Part III   The Philosophy of Geo-Ontologies

**6   Geo-Ontologies: From the Spatial Turn to Geographical
      Taxonomy**   93
6.1   From the Spatial Turn to the Diffusion of Geo-Ontologies   93
6.2   The Problem of Existing Taxonomies   95
6.3   A Geographical Point of View   97
6.4   A Geo-Ontological Tri-Partition   98
        6.4.1   Spatial Geo-Ontologies   99
        6.4.2   Physical (or Natural) Geo-Ontologies   100
        6.4.3   Human Geo-Ontologies   100
        6.4.4   Other Geo-Ontologies   101
6.5   Conclusion   103
References   104

7   Ontological Categories for Geo-Ontologies   107
7.1   The Geo-Ontological Domain   107
7.2   Issues of a Categorial Ontology   108
7.3   On the General Aims of Geo-Ontologies   109
7.4   From the Need of Accessibility to the Common-Sense
        Geography   110
7.5   Ontology Components   111
7.6   Between Formal Ontologies and Ontological Categories   112
7.7   Lowe's Four-Category Ontology   112
7.8   Overlaps and Deviations: Cumpa's Fact-Oriented Ontology   114
7.9   From Components to Contents   114
7.10   Variantism   115
7.11   Two Kinds of Completeness   116
7.12   Completeness and Fundamentality   117
7.13   Philosophical Ontology Versus IT/Computer Ontology   119
References   120

Conclusion   123

Index   125

Part I
Among Computer Science, Philosophy, and Geography: An Ontological Investigation

Chapter 1
From the Philosophies of Geographies to the Applied Ontology of Geography

Abstract This chapter pursues two main goals. The first one is to explore the domain of investigation of applied ontology of geography, by providing an overview of the mutual interactions among the disciplines encompassed in the domain, namely philosophy, geography, and computer science. The second goal is to reveal the disconnections, by delineating some possible strategies designed to increase the interdisciplinary dialogue. In accordance with such goals, Sects. 1.1–1.5 respectively examine the connections between philosophy and geography, philosophy and geographies, philosophical ontologies and geographies, computer science and geographies, and philosophical ontologies and IT/computer ontologies. Section 1.6 acknowledges that the domain of research of applied ontology of geography should include, at least, two different kinds of geography: empirical geography and academic geography. Then, Sects. 1.7 and 1.8 point out that philosophical and IT/computer ontologies are more connected to empirical geography than to academic geography. Finally, Sect. 1.9 explores the reasons why academic geography is currently disconnected to the (philosophical) ontological debate, and then Sects. 1.10 and 1.11 outline some possible strategies to provide a way out of the disconnections.

Keywords Digital geography · Empirical geography · Geography · IT ontology · Naïve realism · Philosophical Ontology · Philosophy of geography · Practices · Representations · Spatial analysis

1.1 On the Connections Between Philosophy and Geography

Getting a complete overview of the possible connections between geography and philosophy might be an extremely ambitious task (Tambassi 2018; Veríssimo Serrão 2018; Tambassi and Tanca 2021). For this reason, instead of providing an exhaustive survey, Tanca (2017, 2018a) rather focuses on some guidelines that describe some current connections between the two disciplines. According to the author, there are essentially four different (and interacting) ways for philosophy and geography to communicate (Fig. 1.1):

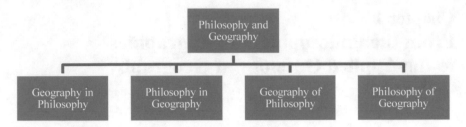

Fig. 1.1 Tanca's (2017, 2018a) connections between philosophy and geography

1.1.1.   geography *in* philosophy;
1.1.2.   philosophy *in* geography;
1.1.3.   geography *of* philosophy;
1.1.4.   philosophy *of* geography.

Dealing with the first two connections means to think in terms of inclusion. More precisely, what is at stake in the *in*-relations is the capacity of each of the two disciplines to open its doors to examples, theories and experiences coming from the other discipline. In this way, we can have geographical notions populating philosophical debates (Kant 1787; Arendt 1955; Benoist 2001; Farinelli 2004; Foucault 2007; Lévy 2021) and philosophical notions infiltrating geographical debates (Harvey 1969; Tuan 1971; Raffestin 1980; Peet 1998; Minca 2001; Church 2011; Lussault 2021). Geography of philosophy is instead centered on those places and/or geographical contexts within which philosophical theories have been developed and philosophers have moved and worked—such as Königsberg for Kant, London for Marx, Berlin for Benjamin, and so forth (Deleuze and Guattari 1994; Holenstein 2004; Rossi and Viano 2004). Finally, philosophy of geography refers to the debate on the nature of geographical knowledge and reality, a debate that includes, among others, ontological, epistemological, social, and cultural groundings of geography as a discipline (Dardel 1952; Farinelli 1992, 2009; Bonesio 2000; Smith and Mark 2001).

As for 1.1.4, philosophy of geography, it might be more appropriate to speak of philosoph*ies* of geograph*ies* in the plural (Fig. 1.2). This shift emphasizes the

Fig. 1.2 Philosophies of geographies

possibility of some mutual connections among different sub-branches of philosophy (ontology, history of philosophy, aesthetics, epistemology, political philosophy, and so on) and of geography (such as human, physical, regional, cultural, and so forth). Accordingly, we have a plurality of sub-disciplines constituting the domain of philosophy of geography, like the phenomenology of human geography, the epistemology of physical geography, the ontology of classic geography, and so on (Talarchek 1977; Nozawa 1996; Inkpen 2005; Azócar Fernández and Buchroithner 2014).

Of course, Tanca's *in/of* connections (Fig. 1.1) are not the only way to sketch an overview of the relations between philosophy and geography. Another chance could be distinguishing the history from the contents of such connections. In the former case, we have a "*history of* the philosophy of geography and/or geographical philosophy", that is, a diachronic reconstruction of the intersections between geographical and philosophical knowledge of a certain historical period (Besse 1998; Elden 2009; Tanca 2012; Veríssimo Serrão 2021). In the latter case, we could refer to *contents* and/or issues (such as space, place, landscape, and so on) at the boundary between philosophical and geographical reasoning (Paquot and Younès 2009; Giubilaro 2016). Obviously, the dichotomy of history and contents does not preclude mixed approaches centered on interdisciplinary contents viewed from some historical perspective.

1.2 Two Triangles

If the overview I have been sketching so far is not enough, one might also consider the connection between geography and philosophy as a ternary relation. According to Tanca (2018b), this would be the result of taking the term "geography" to refer to two different areas of investigation,[1] which correspond to what, in another disciplinary context, the terms "history" and "historiography" respectively denote:

1.2.1. the study of the past in terms of events;
1.2.2. the methods of research of *historians* in developing history as an academic discipline.

Tanca's idea is to apply the distinction between history and historiography to the geographical investigation, so as to split geography into two different parts:

1.2.3. the geographical reality consisting of terrestrial entities and relations among them;
1.2.4. the whole of theories, models and methods aimed at analyzing the geographical reality.

The first part of the geographical investigation takes the name of (small-g) "geography", and the second one the name of (big-G) "Geography". Figure 1.3

---

[1]Similar distinctions, in the geographical debate, can be found in Raffestin (1980); O'Tuathail (1996).

Fig. 1.3 Tanca's (2018b) triangle

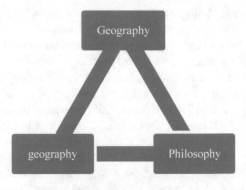

Fig. 1.4 Triangle revisited

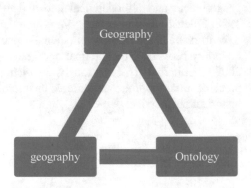

sketches the way in which Tanca summarizes the relationship between philosophy and G/geography: a triangle whose vertexes represent "Geography" in terms of theories, models and methods shared by the community of academic geographers, "geography" conceived as the empirical geographical reality and, finally, "Philosophy".

Insofar as Tanca's triangle is meant to show the mutual connections between philosophy and geography, it should somehow include, according to 1.1.4, all the different philosophical and geographical sub-branches. In order to locate such sub-branches in the triangle, we might consider each vertex as a sort of container for all the different sub-branches of philosophical and geographical investigation. In this sense, focusing on a specific sub-branch would simply mean substituting one vertex, for instance "Geography," with one of its sub-branches, e.g., "human geography".[2]

As this chapter undertakes to explore the domain of investigation of applied ontology of geography, the substitution only concerns the vertex "Philosophy," which should be replaced by the philosophical sub-branch "Ontology" (Fig. 1.4). But what kind of ontology are we referring to?

---

[2]Obviously, this does not mean denying the chance of multiple substitutions; see for example Fig. 1.2.

1.3 Ontologies of Geography

In the last three decades, within the philosophical debate, the ontological research has been characterized by a plurality of hypotheses and methods of investigation, which gave birth to a multiplicity of philosophical inquiries (Varzi 2005; Ferraris 2008). Such a plurality of hypotheses has also involved the relationship between ontology and geography, which currently shows a multifaceted nature. For this reason, it would probably be more accurate to talk about ontologies and geographies in the plural, and then of "ontologies of geographies" (Tambassi 2018).

Indeed, the geographical debate on ontology offers different approaches that have been developed, among others, by Vallega (1995), Berque (2000), Raffestin (2012) and Boria (2013). Regarding the ontological debate on geography, we could say, instead, that it generally involves two main areas of concern (Fig. 1.5). On the one hand, there is the ontological investigation coming from continental philosophy, which may be found in the works of Hacking (2002), Elden (2003), Schatzki (2003), Escobar (2007), Harvey (2007), Dean (2010), Law and Lien (2012), Joronen (2013), Kirsch (2013), Shaw and Meehan (2013), Springer (2013), Whatmore (2013), Blaser (2014), Bryant (2014), Roberts (2014) and Joronen and Häkli (2017). On the other hand, the analytic tradition has developed an ontological reflection on geography that ranges:

1.3.1 from theoretical and technical needs of geographical/IT tools, such as GIS and geo-ontologies (Laurini 2017, 2019; Pesaresi 2017; Couclelis 2019);
1.3.2 to making explicit assumptions and commitments of geography as a discipline, in terms of (classifications of) geographical entities (objects, processes, properties, kinds and so forth), boundaries, and mereo-topological relations

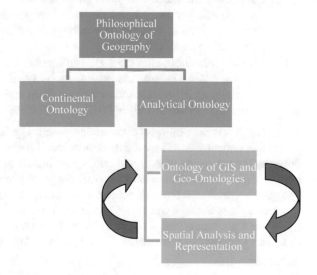

Fig. 1.5 (Philo-)ontological debate on geography

among them and with the space they occupy (Casati et al. 1998; Smith and Mark 1998, 2001; Varzi 2001, 2007, 2016; Smith 2019; Thomasson 2019).

While 1.3.2 is essentially speculative, 1.3.1 is strictly related to computer science. However, we should not conceive 1.3.1 and 1.3.2 as completely independent, but rather as closely intertwined. In particular, 1.3.2 represents (part of) the theoretical investigation behind 1.3.1, which, in turn, constitutes a possible IT application of 1.3.2. This means that, in the analytic tradition, the ontological investigation on geography may be seen as the composition of two connected parts: one more theoretical, one more practical.[3] From now on, we refer to the whole of 1.3.1 and 1.3.2 by the name of "applied ontology of geography" [AOG].

1.4   Digital Geographies

Moreover, we could add that, as AOG deals with geographical/IT tools such as GIS and geo-ontologies (see 1.3.1), it (also) concerns the digital dimension of geographical investigation. According to Sui and Morrill (2004), that dimension has progressively affected the practice of geography, transforming both the academic geography and the geography of the world—in particular, the ways in which geographic knowledge is constructed, debated and communicated.

In order to avoid thinking about a radical rupture with the geographical theories and praxes of the past, Ash et al. (2018, p. 1) suggest describing such a transformation with the locution "digital turn" which is aimed at:

1.4.1. capturing the ways in which there has been a significant turn to the digital as both object and subject of geographical inquiry;

1.4.2. signaling the modalities through which the digital (more precisely, software packages and digital devices) has pervasively inflected and mediated geographic thought, research, practice, and knowledge production, generating many geographies;

1.4.3. highlighting how, after the digital turn, notions such as experience spatiality, space, place, nature, landscape, mobility, and environment have been partially reshaped (Castells 1996; Graham and Marvin 2001; Elwood and Leszczynski 2013; Wilson 2012; Rose et al. 2014).

Although it is not (entirely) clear whether digital geography should be conceived as an independent area of investigation with specific aims and points of view, for our purpose, the reflection of Ash and colleagues is still interesting for their threefold categorization of the relationship between geography and digitality, which includes geographies produced *through*, produced *by* and *of* the digital.

Geographies *through* the digital generally refer to the fact that digital has long figured as a prominent site, mode and object of/for knowledge production, especially in human geography (Rose 2015). For this reason, "the digital has been engaged to

---

[3] Such parts are respectively analyzed in Chaps. 3–5 and Chaps. 6–7.

actualize heterodox epistemologies in the service of producing geographic knowledge, while simultaneously being the subject of epistemological critique" (Ash et al. 2018, p. 3). Geographies produced *by* the digital represent studies that examine how the digital is mediating, augmenting (through digital interfaces) and integrating (with various kinds of georeferenced and real-time data) the production of space, transforming socio-spatial relations and generating new kinds of spatiality. Finally, geographies *of* the digital apply geographical ideas and methodologies in order to chart and map the digital, which includes the spatial and technical realm of communication and interaction (the internet/cyberspace, virtual worlds, digital games), as well as associated socio-technical assemblages of production.

1.5 Big-O Ontology Versus Small-o Ontologies

Now, if we want to clarify what the term "ontology" precisely means in a context of philosophy, geography and digitality, we should start by saying that AOG considers two different options, which Guarino and Giaretta (1995) identify with (big-O) Ontology and (small-o) ontologies.

(Big-O) Ontology comes from the analytical tradition of the philosophical investigation, which refers "ontology" to the discipline concerned with the question of "what entities exist". This means we need to outline, at a high level of abstraction, a complete inventory of reality by specifying its hierarchical and categorial structure (Varzi 2005). Moreover, Ontology is often associated with questions about existence, identity, reality, cosmos, minds, ideas, consciousness, space and time, language and truth (Kavouras and Kokla 2011; Couclelis 2019).

(Small-o) ontologies are instead the result of the work of IT/computer scientists, which focus on the technical problems of knowledge sharing and interoperability among databases. In such a context, ontology is generally conceived as an explicit (and sometimes partial) specification of a shared conceptualization that is formalized in a logical theory (Gruber 1993; Guarino and Giaretta 1995; Borst 1997; Studer et al. 1998; Tambassi and Magro 2015). Along with this definition, Gruber adds that any IT/computer ontology should also

1.5.1.   contain a description of the concepts and relationships that can exist for an agent or a community of agents;
1.5.2.   constrain the possible interpretations for the defined terms.

Thus, an ontology is, by definition, "relative to specific agents facing specific tasks individually or as a community (e.g., a scientific or professional community), and its role is to pin down the fluid meanings of terms in ways that serve the needs of that individual or community at some specific time" (Couclelis 2019, p. 5). To sum up, ontologies in IT/computer science are often referred to as "small-o ontologies", because they are aimed at describing the finite artificial worlds defined by these databases. The use of the plural in "ontologies" indicates a multiplicity of ways to represent such worlds as well as a plurality of ontologies, each developed for a specific use, context, and purpose (Adams and Janowicz 2011; Janowicz and Hitzler 2012). Finally, it is worth noting that the development of the Semantic Web has been generating additional difficulties for the task of IT/computer ontology construction, "because Web knowledge has no primitives, no core, no fundamental categories, no fixed structure, and is heavily context-dependent" (Couclelis 2019, p. 4).[4]

1.6   The Geo-Ontological Circle

In line with the terminology of the previous sections, we could consider the domain of AOG as composed of four different areas of research: (big-G) Geography, (small-g) geography, (big-O) Ontology and (small-o) ontology.

(Big-G) Geography indicates the academic discipline, which includes the totality of theories, models and methods shared and discussed by the community of professional geographers. However, as Wright (1947) suggests, geographical knowledge is not the monopoly of professional geographers. Indeed, "the lexical and semantic field defined by their discourses is more restricted and selective than the broader scope of everything that can be said about geographic reality as we know it" (Tanca 2017, p. 2). Accordingly, if academic (big-G) Geography represents a small core area, there is also a much broader peripheral zone that includes the informal geography contained in non-scientific works, such as books of travel, magazines, newspapers, fiction poetry, canvas, and so forth (Wright 1947, p. 10). Such informal (small-g) geography refers to the empirical geographical reality, that is, the sum of the *things* on Earth and of the processes going on among them.

From the ontological side, (big-O) Ontology concerns the philosophical (analytical) domain, where the term indicates the study of "what there is/exists". Its main aim is to provide an exhaustive inventory of reality, in terms of entities (facts, objects, relations, properties and so forth) and structure. Geographically speaking, the scope of Ontology is to analyze the geographical world so to outline what kinds of geographical entities exist and how they can be classified and related in a system that gathers them together. Finally, (small-o) ontology refers to IT/computer science that conceives ontologies as formal specifications of shared conceptualizations. Such specifications define the general concepts of geographical entities and relations, in order to describe domains of geographical application or, at least, some of their very specific sub-areas.

Figure 1.6 summarizes the domain of investigation of AOG, which might be represented as a geo-ontological circle.

---

[4]For an analysis on the different kinds of ontologies behind AOG, see Chap. 2.

Fig. 1.6 Geo-ontological circle

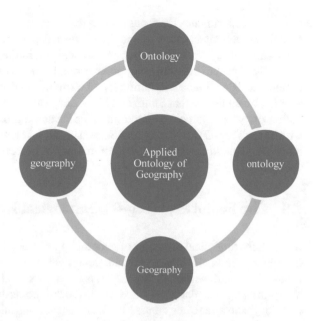

1.7   (Small-o) Ontological Interdisciplinary Connections

Now, if Geography, geography, Ontology and ontology cover the domain of AOG, what can we say about the relations among them?

In answering this question, the starting point of Tanca (2017) is a quote from Deleuze, according to which "the encounter between two disciplines does not take place when one begins to reflect on the other, but when one discipline realizes that it has to resolve, for itself and by its own means, a problem similar to one confronted by the other" (Deleuze 2014, p. 387). And this happens for ontology and Ontology. Both share the task of providing an (exhaustive) inventory of entities (and relations among them), but they differ in the kind of inventory they have to supply, the tools they use, and the domains they have to systematize. On this basis, the *encounter* between ontology and Ontology has quickly transformed into a mutual relation, which came with, at least in some cases, a set of common solutions (Tambassi and Magro 2015; Tambassi 2021).

The connection between ontology and geography, instead, is grounded on the notion of (geographical) conceptualization, which is endorsed by both disciplines. About this notion, we should mention the experiments of Smith and Mark (2001),[5] aimed at:

1.7.1. analyzing how non-expert subjects conceptualize geospatial phenomena;
1.7.2. providing a first approximation of the basic noun lexicon for formal geographical ontologies (our small-o ontologies).

---

[5]The results of these experiments are discussed in Chap. 2.

In such experiments, the focus on non-expert conceptualizations indicates a theoretical choice that is more oriented to informal geography than to the academic one. And within informal geography, the subjects in the experiments (see Chap.2) primarily conceptualize the geographical reality in terms of concrete geographical things, that is, in Tanca's terminology, in terms of small-g geography. However, we should also underline that Smith and Mark link scientific/academic geography to the notion of "what can be portrayed on maps". Such a link is not totally overlapping with Tanca's big-G Geography (see Sects. 1.9 and 1.11), which also refers to theories, models and methods shared by the community of professional geographers.

1.8 Five Features of (Big-O) Ontological Knowledge

In 1.3.2, the analytical ontology of geography has been defined as the philosophical discipline specifically concerned with geographical entities, boundaries, and mereotopological relations among them. As it is mainly focused on describing the different *things* that populate Earth's surface, Tanca (2018a) holds that the discipline (the big-O Ontology) shows a strong connection with conceptualizations coming from (small-g) geography. Moreover, according to Tanca, the analytical ontology of geography is characterized by, at least, five different theoretical assumptions:

1.8.1.   naïve realism;
1.8.2.   implicit epistemology;
1.8.3.   primacy of sight;
1.8.4.   metaphysics of objects;
1.8.5.   neglect of temporal dimension.

Naïve realism, in such a context, combines two different theses: (1) reality is ultimately mind-independent (or independent from our way to know and describe it); (2) there is in principle no obstacle to our knowing at least something about reality as it is in itself (Lowe 2002). This means that geographical properties are ultimately *in rebus*, and that our knowledge of the geographic world is faithful to the world itself (regardless of the fact that we can *actually* know everything about the geographic world) and can also be obtained through the use (or the development) of specific tools of investigation (Orain 2009).

Implicit epistemology refers to the fact that the whole of enquiry procedures, methods and tools shared by the community of academic geographers is neither questioned nor critically discussed. This also applies to the validity of their conceptual schemas, as well as to the social, political, and cultural value of their descriptions of the geographical world. Such an implicit epistemology is also reflected in the primacy of sight as a means of (privileged) access to the geographical reality. Surely, the geographical sight is something that requires some sort of training. However, once the training has been honed, the sight enables us to focus on details and connections which are invisible for others. And where the sight alone is not enough, maps and other visual geographical tools can help us in knowing new "things" on/of the geographical world.

Metaphysics of objects means that the geographical world is ultimately the sum of concrete and persistent (geographical) objects, which are characterized by distinctive properties and linked together by spatial and geographical relations. Such (geographical) objects are generally distinguished in two main realms: human and physical. The former includes items such as artifacts and entities like states, provinces and regions that depend on our delineating or conceptualizing activities. The latter comprehends objects like rivers, hills, and oceans, which exist independently of all human cognitive acts and demarcations (Smith 1995, 2019; Casati et al. 1998; Galton 2003).

Finally, the neglect of the temporal dimension records that geographical objects are generally conceived in terms of current/actual existence. This means ignoring the diachronic persistence of such objects and taking a snapshot of the geographical world without taking into account any kind of transformation.

1.9   (Big-G) Geographical Disconnections

Sections 1.7–1.8 have focused on the interdisciplinary connections involving Ontology, ontology, and geography. Now, what could we say about Geography? More precisely, how is Geography connected to the other disciplines of the geo-ontological circle (Fig. 1.6)?

Concerning the connection between Geography and ontology as a branch of digital computation (see Sect. 1.5), Ash and colleagues observe that,

> [academic] geographers are uniquely placed to *interrogate* the materialities of digital computation in innovative ways. Geographers' theorizations of space, time and relationality can be fruitfully developed to consider how digital computation and its associated objects are both singular things, with particular capacities, that also create shared space times for both other technical objects and the humans who use those objects. This calls for further attention to be given to the work that non-human infrastructures perform that always exceeds the technical parameters of their design. (Ash et al. 2018, p. 12)

According to Tanca, one of the reasons why the *interrogation* has not yet happened consists in the "[academic] geographers' lack of interest or fear of going beyond the limits of their disciplinary field" (Tanca 2017, p. 3). And such a lack of interest is not just about the connection between Geography and ontology, but also about the relation between Geography and Ontology, and between Geography and geography. This means, there is a general lack of dialogue between Geography and other disciplines of the geo-ontological domain. The lack of dialogue has made (small-g) geography progressively fill the void left by Geography, to the extent that it has become the geographical reference point for the other disciplines.

> The other side of the coin of this reasoning is that, due to this ambiguity, we often witness a separation, a lack of synchrony [...] between what happens 'inside' and 'outside' geography, i.e. the community of scholars who acknowledge certain theories and scientific archetypes and the research tools linked to them. Very often, those who deal with 'geographic' themes

Fig. 1.7 Interdisciplinary
relations

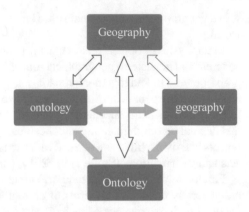

do not waste their time by reading what we geographers write, and therefore do not know about the most recent developments within the discipline. To the eyes of a critic, geography as it is practised outside geography appears to be occasionally burdened by anachronisms, by statements that we can easily deem naïve or obsolete, by subject matters which are presented as they were brand new but have actually been discussed and rediscussed for twenty or thirty years in periodicals and conventions (Tanca 2017, p. 3).

Figure 1.7 represents connections and disconnections among the different disciplines covering the geo-ontological domain. Light blue arrows indicate connections, whereas white arrows stand for (current) disconnections.

Ash et al. (2018) underline that the proliferation, commercialization and popularization of geographical technologies is engendering the flourishing of spatial ontologies and epistemologies—and then of the debate on AOG. Their suggestion is thus that academic geographers adopt and embrace an epistemological, ontological and methodological openness in their engagements with the digital in general and with the AOG in particular. Obviously, this suggestion does not prevent applied ontologists of geography from enriching their research with methods and tools coming from big-G Geography.

But what are academic geographers and applied ontologists of geography specifically supposed to do? In Sects. 1.10 and 1.11, we sketch some possible strategies for developing an interdisciplinary dialogue between Geography and the other disciplines of the geo-ontological circle.

1.10 Representations and Practices

In dealing with an area of research that ranges from human to physical and technical analysis (Sala 2009), Geography shows a multifaceted (and hybrid) nature, which also affects its Ontological background. For this reason, Tanca (2018a) suggests, it would be reductive to think about the connection between Ontology and Geography only in terms of "things" and/or "objects". In other words, an Ontological reflection on Geography based on the notion of mirror of nature (see Sect. 1.8) is not enough for explaining the Ontological background of Geography. Therefore, an Ontological reflection should also include the other *joints* of Geographical investigation: i.e., representations and practices.

Following Tanca, the term "representations" includes notions such as perceptions, experience spaces, texts, images and so forth. In the context of this representative joint, Tanca highlights the constructivist assumption that the geographical world takes shape *within* a network of symbols and representations. The assumption combines two different ideas. The former is that the geographical world does not pre-exist our cognitive schemes—in other words, it is mind-dependent. The latter is that our knowledge of the geographical world does not put us in direct contact with the world *as it is*. In this sense, geographical knowledge is the activity of creating and producing meanings. This signifies acknowledging the performative role of

1.10.1.   language, which does not represent mimetically the geographical world;
1.10.2.   the subject in producing meanings through language;
1.10.3.   social and cultural contexts, within which subjects live.

Practices are not less important than representations and, according to Tanca, mainly refer to *non-representational theories*. Such theories are based on the assumption that the dualism between "things" (naïve realism) and "representations" (social constructivism) does not fully explain the complexity of the geographical world. The idea is thus to integrate things and representations by making explicit the dynamic and processual character of geographical reality and the modality of production of sense in terms of practices, procedures and habits. Theoretically speaking, non-representational theories recognize, at least, three different sources of inspiration: phenomenology, new vitalism and post-structuralism. The mixture results in close attention to bodies and expressive components of experience (emotions, affections, memories), which are integrated with things and representations. But the mixture also results in the primacy of practices, performances, thought-in-action and action-in-context, that is, the recognition of the inseparability of subjects and the geographical world as necessarily influencing each other.

1.11   From Applied o/Ontology of Geography to Applied o/Ontology of Geography

Section 1.10 suggests a strategy aimed at integrating (part of) the Geographical debate within the domain of AOG. More precisely, the idea is to consider the Ontological background of Geography not only in terms of geographical things, but as a whole composed of geographical things, representations and practices. Such a whole is meant, in principle, to extend the geo-ontological conceptualization behind AOG—provided by (small-g) geography——through the inclusion of ontological reflections coming from (big-G) Geography.

Still, this is not the only strategy. Another chance might be to critically analyze the time dimension. As stated in Sect. 1.8, the neglect of time implies that, in AOG, geographical objects are generally conceived just in terms of current existence, which corresponds to a snapshot of their present or of their existence in a specific period of time. Here, I do not want to claim that neglecting the diachronic dimension is an actual problem for all positions of the current debate in AOG, forasmuch as some geo-ontologies (see Chap. 6) and geo-ontological investigations (Varzi 2019) incorporate this dimension in terms of evolution and change of geographical entities over time. Rather, I would like to refer to a general neglect of the temporal dimension within (small-G) geography, which prevents us from considering the geographical investigation as an ongoing process that involves paradigm shifts, and modifications of geographical perspectives, conceptualizations, tools, branches of investigation and so forth. It is in such a meta-theoretical sense that stopping the neglect of the temporal dimension is crucial: asking right now what a geographical object is, is different from asking the same question 1000 years ago, since the notion of geographical object (Fig. 1.9) as well as Geography as a discipline (Fig. 1.8) have been changing constantly.

Finally, since the previous strategies were mainly focused on the introduction of Geographical reflections within a domain that is also composed of geography, Ontology and ontology, we might ask how the reverse process could be obtained, that is, the introduction of reflections coming from geography, Ontology and ontology into the Geographical debate. Ash et al. (2018) claim that only professional geographers have the power to do that. If so, we might also observe that the introduction of AOG's reflections within Geography should not mean subordinating AOG to Geography, but enriching AOG's conceptualizations Geographically, by starting from the (small-g) geography that AOG currently presupposes.

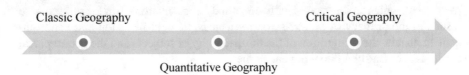

Fig. 1.8 Geography as a process

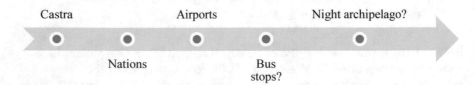

Fig. 1.9 Geographical objects as a process

References

Adams BJ, Janowicz K (2011) Constructing geo-ontologies by reification of observation data. ACM SIG Spatial GIS. ACM Press, Chicago, pp 309–318

Arendt A (1955) The Hannah Arendt Papers at the Library of Congress (https://memory.loc.gov/ammem/arendthtml/arendthome.html). Subject File, 1949–1975, n.d.: Courses—University of California, Berkeley, Calif. – 'History of Political Theory', lectures—Tocqueville, Alexis de, and Karl Marx, and conclusion (024090–024091)

Ash J, Kitchin R, Leszczynski A (2018) Digital turn, digital geographies? Prog Hum Geogr 42(1):25–43

Azócar Fernández PI, Buchroithner MF (2014) Paradigms in cartography. An epistemological. Review of the 20th and 21st Centuries. Springer, Heidelberg

Benoist Y (2001). In: In Benoist J, Merlini F (eds) quoi la géographie peut-elle importer à la philosophie? Historicité et spatialité. Le problème de l'espace dans la pensée contemporaine. Vrin, Paris, pp 221–247

Berque A (2000) Mediance: de milieux en paysages. Belin, Paris

Besse J-M (1998) La philosophie et la géographie. In: Mattei J-F (ed) Encyclopédie philosophique universelle, IV. PUF, Paris, pp 2550–2566

Blaser M (2014) Ontology and indigeneity: on the political ontology of heterogenous assemblages. Cult Geogr 21(1):49–58

Bonesio L (2000) Terra, singolarità, paesaggi. In: Bonesio L (ed) Orizzonti della geofilosofia. Terra e luoghi nell'epoca della mondializzazione. Arianna, Casalecchio, pp 5–25

Boria E (2013) Genealogie intellettuali e discontinuità nazionali nella storia della cartografia. Bollettino Della Società Geografica Italiana 6(3):443–460

Borst WN (1997) Construction of engineering ontologies, centre for telematica and information technology. University of Twente, Enschede

Bryant LR (2014) Onto-cartography. An ontology of machines and media. Edinburgh University Press, Edinburgh

Casati R, Smith B, Varzi AC (1998) Ontological tools for geographic representation. In: Guarino N (ed) Formal ontology in information systems. IOS Press, Amsterdam, pp 77–85

Castells M (1996) The rise of the network society. Blackwell, Oxford

Church M (2011) Immanuel Kant and the emergence of modern geography. In: Elden S, Mendieta E (eds) Reading Kant's geography. State University of New York Press, Albany, NY, pp 19–46

Couclelis H (2019) Unpacking the "I" in GIS: information, ontology, and the geographic world. In: Tambassi T (ed) The Philosophy of GIS. Springer, Cham, pp 3–24

Dardel E (1952) L'Homme et la Terre: nature de la réalité géographique. PUF, Paris

Dean M (2010) Governmentality: power and rule in modern society. SAGE, London

Deleuze G (2014) The brain is the screen. Interview with Gilles Deleuze. Diagonal thoughts. http://www.diagonalthoughts.com/?p=2252

Deleuze G, Guattari F (1994) What Is philosophy? Columbia University Press, New York

Elden S (2003) Reading genealogy as historical ontology. In: Milchman A, Rosenberg A (eds) Foucault and Heidegger: critical encounters. University of Minnesota Press, Minneapolis

Elden S (2009) Philosophy and human geography. In: Kitchin R, Thrift N (eds) International encyclopaedia of human geography. Elsevier, Oxford, pp 145–150

Elwood S, Leszczynski A (2013) New spatial media, new knowledge politics. Trans Inst Br Geogr 38:544–559

Escobar A (2007) The 'ontological turn' in social theory. Trans Inst Br Geogr 32(1):106–111

Farinelli F (1992) I segni del mondo: immagine cartografica e discorso geografico in età moderna. La Nuova Italia, Firenze

Farinelli F (2004) Experimentum mundi. Introduction to I. Kant, Geografia fisica. Riproduzione anastatica dell'edizione Silvestri 1807–1811. Leading Edizioni, Bergamo, pp I–XXIX

Farinelli F (2009) La crisi della ragione cartografica. Einaudi, Torino

Ferraris M (ed) (2008) Storia dell'ontologia. Bompiani, Milano

Foucault M (2007) The meshes of power. In: Crampton JW, Elden S (eds) Space, knowledge and power. Foucault and geography. Ashgate, London, pp 153–162

Galton A (2003) On the ontological status of geographical boundaries. In: Duckham M, Goodchild MG, Worboys MF (eds) Foundation of geographic information science. Taylor & Francis, London-New York, pp 151–171

Giubilaro C (2016) Corpi, spazi, movimenti: per una geografia critica della dislocazione. UNICOPLI, Milano

Graham S, Marvin S (2001) Splintering urbanism: networked infrastructures, technological mobilities and the urban condition. Routledge, London

Gruber TR (1993) A translational approach to portable ontology specifications. Knowl Acquis 5(2):199–220

Guarino N, Giaretta P (1995) Ontologies and knowledge bases—towards a terminological clarification. In:Mars NJ (ed) *Towards very large knowledge bases—knowledge building and knowledge sharing*. IOS Press, Amsterdam, pp 25–32

Hacking I (2002) Historical ontology. Harvard University Press, Cambridge, MA

Harvey D (1969) Explanation in geography. E. Arnold, London

Harvey D (2007) The Kantian roots of Foucault's dilemmas. In: Crampton JW, Elden S (eds) Space, knowledge and power: foucault and geography. Ashgate, Hampshire

Holenstein E (2004) Philosophie-Atlas: Orte und Wege des Denkens. Ammann, Zürich

Inkpen R (2005) Science, Philosophy and Physical Geography. Routledge, London-New York

Janowicz K, Hitzler P (2012) The digital earth as knowledge engine. Editorial. Semant Web 1(1):1–10

Joronen M (2013) Heidegger, event and the ontological politics of the site. Trans Inst Br Geogr 38(4):627–638

Joronen M, Häkli J (2016) Politicizing ontology. Prog Hum Geogr 41(5):561–579

Kant I (1787) Critique of pure reason. Cambridge University Press, Cambridge (1998)

Kavouras M, Kokla M (2011) Geographic ontologies and society. In: Nyerges TL, Couclelis H, McMaster R (eds) The SAGE handbook of GIS and society. SAGE, London, pp 46–66

Kirsch S (2013) Cultural geography I. Materialist turns. Prog Hum Geogr 37(3):433–441

Laurini R (2017) Geographic knowledge infrastructure: Applications to territorial intelligence and smart cities. ISTE-Elsevier, London

Laurini R (2019) Some philosophical issues regarding geometric modeling for geographic information and knowledge systems. In: Tambassi T (ed) The philosophy of GIS. Springer, Cham, pp 25–50

Law J, Lien ME (2012) Slippery: field notes in empirical ontology. Soc Stud Sci 43(3):363–378

Lévy J (2021) Transmutations. Novel encounters between philosophy and social theory of space. In: Tambassi T, Tanca M (eds) The philosophy of geography. Springer, Cham

Lowe EJ (2002) A survey of metaphysics. Oxford University Press, Oxford

Lussault M (2021) Spatiality first. In Tambassi T, Tanca M (eds), The Philosophy of Geography. Springer, Cham

Minca C (2001) Introduzione alla geografia postmoderna. CEDAM, Padova

Nozawa H (ed) (1996) Social theory and geographical thought. Kyushu University, Fukuoka

Orain O (2009) De plain-pied dans le monde. Ecriture et réalisme dans la géographie française au XXe siècle. L'Harmattan, Paris

O'Tuathail G (1996) Critical geopolitics: the politics of writing global space. Routledge, London

Paquot T, Younès C (eds) (2009) Le territoire des philosophes. Lieu et espace dans la pensée du XXe siècle. la Découverte, Paris

Peet R (1998) Modern geographical thought. Blackwell, Oxford

Pesaresi C (2017) Applicazioni GIS. Principi metodologici e linee di ricerca. Esercitazioni ed esemplificazioni guida. Utet, Torino

Raffestin C (1980) Pour une géographie du pouvoir. Librairies Techniques, Paris

Raffestin C (2012) Space, territory, and territoriality. Environ Plan D: Soc Space 30(1):121–141

Roberts T (2014) From things to events: whitehead and the materiality. Environ Plan D: Soc Space 32:968–983

Rose G (2015) Rethinking the geographies of cultural 'objects' through digital technologies: interface, network and friction. Prog Hum Geogr 40(3):334–351

Rose G, Degen M, Melhuish C (2014) Networks, interfaces, and computer-generated images: learning from digital visualisations of urban redevelopment projects. Environ Plan A 32:386–403

Rossi P, Viano CA (eds) (2004) Le città filosofiche: per una geografia della cultura filosofica italiana del Novecento. il Mulino, Bologna

Sala M (2009) Geography. In: Sala M (ed) Geography. Encyclopedia of life support systems. EOLSS Publisher, Oxford, pp 1–56

Schatzki T (2003) A New societist social ontology. Philos Soc Sci 33(2):174–202

Shaw IG, Meehan K (2013) Force-full: power, politics and object-oriented philosophy. Area 45(2):216–222

Smith B (1995) On drawing lines on a map. In: Frank AU, Kuhn W, Mark DM (eds) Spatial information theory. Proceedings of COSIT '95, Springer, Berlin-Heidelberg-Vienna-New York-London-Tokyo, pp 475–484

Smith B (2019) Drawing boundaries. In: Tambassi T (ed) The Philosophy of GIS. Springer, Cham, pp 137–158

Smith B, Mark DM (1998) Ontology and geographic kinds. In: Poiker TK, Chrisman N (eds) Proceedings of the eighth international symposium on spatial data handling (Burnaby, British Columbia, International Geographical Union), pp 308–320

Smith B, Mark DM (2001) Geographical categories: an ontological investigation. Int J Geogr Inf Sci 15(7):591–612

Springer S (2013) Human geography without hierarchy. Prog Hum Geogr 38(3):402–419

Studer R, Benjamins VR, Fensel D (1998) Knowledge engineering: principles and methods. IEEE Trans Data Knowl Eng 25(1–2):161–197

Sui D, Morrill R (2004) Computers and geography: from automated geography to digital earth. In: Brunn SD, Cutter SL, Harrington JW (eds) Geography and technology. Springer, New York, pp 81–108

Talarchek GM (1977) Phenomenology as a new paradigm in human geography. Syracuse University, Department of Geography, Discussion Paper Series, p 39

Tambassi T (2018) Introduzione. Filosofia della Geografia. Semestrale di Studi e Ricerche di Geografia 30(2):9–14

Tambassi T (2021) Intrecci meta-ontologici tra filosofia e informatica. Giornale di metafisica, in press

Tambassi T, Magro D (2015) Ontologie informatiche della geografia. Una sistematizzazione del dibattito contemporaneo. Rivista Di Estetica 58:191–205

Tambassi T, Tanca M (2021) The philosophy of geography. Springer, Cham

Tanca M (2012) Geografia e filosofia. Materiali di lavoro. Franco Angeli, Milano

Tanca M (2017) Incongruent counterparts. Four possible ways of interaction between geography and philosophy. J Interdiscip History Ideas 6(12):1–38

Tanca M (2018a) Cose, rappresentazioni, pratiche: uno sguardo sull'ontologia ibrida della Geografia. Bollettino della Società Geografica Italiana 14, 1(1):5–17

Tanca M (2018b) Geografia e filosofia: istruzioni per l'uso. Semestrale Di Studi e Ricerche Di Geografia 30(2):13–27

Thomasson AL (2019) Geographic objects and the science of Geography. In: Tambassi T (eds) The philosophy of GIS. Springer geography. Springer, Cham

Tuan Y (1971) Geography, phenomenology, and the study of human nature. Can Geogr 15(3):181–192

Vallega A (1995) La regione, sistema territoriale sostenibile: compendio di geografia regionale sistematica. Mursia, Milano

Varzi AC (ed) (2001) The philosophy of geography. Special issue of Topoi 20(2)

Varzi AC (2005) Ontologia. Rome-Bari, Laterza

Varzi AC (2007) Spatial reasoning and ontology: parts, wholes and location. In: Aiello M, Pratt-Hartmann I, van Benthem J (eds) Handbook of spatial logics. Springer, Berlin, pp 945–1038
Varzi AC (2016) On drawing lines across the board. In: Zaibert L (ed) The theory and practice of ontology. Palgrave Macmillian, London, pp 45–78
Varzi AC (2019) What is a city? Topoi, 1–10. https://doi.org/10.1007/s11245-019-09647-4
Verissimo Serrão A (2018) Una mirada de la Filosofía sobre la Geografía. Dos visiones del mundo, una comunidad de problemas. Semestrale di Studi e Ricerche di Geografia 30(2):31–41
Verissimo Serrão A (2021) Converting the Earth into a Dwelling Place. The Ethics of Nature by Ludwig Feuerbach. In: Tambassi T, Tanca M (eds) The philosophy of geography. Springer, Cham
Whatmore S (2013) Earthly powers and affective environments: an ontological politics of flood risk. Theory Cult Soc 30(7–8):33–50
Wilson MW (2012) Location-based services, conspicuous mobility, and the location-aware future. Geoforum 43:1266–1275
Wright JK (1947) Terrae incognitae: the place of the imagination in geography. Ann Assoc Am Geogr 37(1):1–15

Chapter 2
The Ontological Background

Abstract Chapter 1 showed that applied ontology of geography includes four different areas of research, namely: IT/computer ontology, philosophical ontology, empirical geography, and academic geography. In this chapter, we go deep into the analysis of the (kinds of) ontologies behind applied ontology of geography. Sections 2.1–2.3 explores the domain of IT/computer science, within which ontologies are generally conceived as explicit specifications of shared conceptualizations. Sections 2.4 and 2.5 deal with philosophy, more precisely, with the recent proliferation of ontological research in the analytic debate. Finally, Sects. 2.6–2.10 discuss the advancements of ontology of geography, that is that part of philosophical ontology mainly focused on: (1) establishing what geographical entities exist, (2) developing a theory of spatial representation, and (3) explaining how the geographic descriptions of reality emerging from common sense can be combined with those derived from academic geography. This preliminary analysis is meant to provide a helpful framework showing the importance of geographical common-sense conceptualizations and defining how non-expert subjects conceptualize geospatial phenomena in response to a series of different geographical phrased elicitations.

Keywords Common-sense conceptualizations · IT/computer ontologies · Metaphysics · Ontology of geography · Ontological turn · Philosophical ontology · Regional ontologies

2.1 IT/Computer Ontologies and Semantic Web

In the wake of the birth and recent developments of the Semantic Web, computer science has been granting new attention to ontologies. The main idea behind the Semantic Web is extending the classical Web to a "Web of Data" in which the meaning of such data, their semantics, and the information resources designed and built for human fruition are *understandable* also for artificial agents (e.g., software applications).[1]

---

[1] Goy and Magro (2015), p. 7463.

© The Author(s), under exclusive license to Springer Nature Switzerland AG 2021
T. Tambassi, *The Philosophy of Geo-Ontologies*, SpringerBriefs in Geography,
https://doi.org/10.1007/978-3-030-78145-3_2

Enabling software applications to understand at least some meanings of data is aimed to support the design and the implementation of smart applications, and to facilitate data *communication, sharing, inference, interoperability, aggregation,* and *integration* on the Web. The most important tool for describing data meaning are ontologies (the small-o ontologies of Chap. 1), which thus play a major role in Semantic Web architecture.[2]

Ontologies serve especially to support *communication* between:

2.1.1.   human beings,
2.1.2.   human beings and software systems,
2.1.3.   software systems.

In the case of 2.1.1, ontologies are informal, or semi-formal, representations, which are used by companies to enable concepts and *standard terminology sharing,* within a single company or between different companies.

In the case of 2.1.2, ontologies enhance information access. Examples of this use are the so-called "semantic portals" which link contents to concepts and to relationships (explicitly represented in the portal) so as to improve the *aggregation* and *integration* of such contents.

In the case of 2.1.3, the importance of an explicit use of semantic representations has been acknowledged also in the field of information search on the Web (or in large information repositories) in virtue of the role it has in advancing search service performance ("semantic search").

> This means building search engines with user interfaces presenting the underlying "semantic knowledge" to users, thus enabling her/him to express, refine, expand her/his query on the basis of this knowledge. [...] The *communication* between software applications usually consists in exchanging messages. In order to enable such an exchange, the involved applications must share, at least, the encoding format, the information representation syntax, and some communication protocols. However, in order to "understand" the information produced by another application, a *shared* reference conceptualization is needed. In many cases, this level is not explicitly represented, but it is hardcoded within the application code. [...] [Moreover] ontologies could be used to achieve (at least a partial) semantic *interoperability,* by sharing the same ontologies, by translating internal representations of the exchanged contents in terms of a shared ontology, by explicitly specifying a mapping between the different ontologies adopted by the involved applications or between them and a shared ontology. [...] Last, but not least, ontologies expressed in logical languages enable automatic *inferences,* aiming at making explicit knowledge that is implicit, either in ontologies or in the datasets characterized in terms of them (Goy and Magro 2015).

---

[2]The W3C16 has defined and still maintains several standards for the Semantic Web, among which there are two formal languages for ontology specification: RDF Schema (RDFS) and Web Ontology Language (OWL). RDFS is a specification enabling the definition of RDF vocabularies (a general-purpose language for representing information on the Web). OWL is an ontology language for the Semantic Web with formally defined meaning, usually exploited to express ontologies.

2.2 Some Definitions of IT/Computer Ontology

Unlike the philosophical debate (see Sect. 2.5), in IT/computer science, the term "ontology" does not refer to a global and unambiguous characterization of reality (describing its fundamental structures), but to the representation of a particular viewpoint on a portion of reality.[3] This means that, in IT/computer science, there can be many "ontologies", often partial, referring to some aspects or parts of existence and including any possible area of interest.

Moreover, computer scientists do not agree on the meaning of "ontology". Indeed, the literature has provided different definitions of the term, some of which are listed here below.[4]

2.2.1. Neches et al. (1991): an ontology defines the basic terms and relations comprising the vocabulary of a topic area as well as the rules for combining terms and relations to define extensions to the vocabulary;

2.2.2. Gruber (1993): an ontology is an explicit specification of a conceptualization;

2.2.3. Guarino and Giaretta (1995): ontology is a logical theory which gives an explicit, partial account of a conceptualization;

2.2.4. Bernaras et al. (1996): an ontology provides the means for describing explicitly the conceptualization behind the knowledge represented in a knowledge base;

2.2.5. Borst (1997): ontologies are formal specifications of shared conceptualizations;

2.2.6. Swartout et al. (1997): ontology is defined as a hierarchically structured set of terms for describing a domain that can be used as a skeletal foundation for a knowledge base;

2.2.7. Studer et al. (1998): an ontology is a formal, explicit specification of a shared conceptualization;

2.2.8. Guarino (1998); an ontology is a logical theory accounting for the intended meaning of a formal vocabulary, i.e., its ontological commitment to a particular conceptualization of the world. The intended models of a logical language using such a vocabulary are constrained by its ontological commitment. An ontology indirectly reflects this commitment (and the underlying conceptualization) by approximating these intended models;

2.2.9. Uschold and Jasper (1999): an ontology may take a variety of forms, but it necessarily includes a vocabulary of terms and some specification of their meaning. This includes definitions and indication of how concepts are interrelated which collectively impose a structure on the domain and constrain the possible interpretations of terms. An ontology is virtually

---

[3]The term "reality" is used in a broad sense, which includes, for example, physical entities, counterfactual ones, imaginary entities, and so forth (Goy and Magro 2015, p. 7457).

[4]Other lists of definitions of ontology in computer science can be found in Gómez-Pérez et al. (2004), Bullinger (2008), Jaziri and Gargouri (2010).

always the manifestation of a shared understanding of a domain that is agreed between a number of agents. Such agreement facilitates accurate and effective communication of meaning, which in turn leads to other benefits such as interoperability, reuse and sharing;

2.2.10.   Sowa (2000): the subject of ontology is the study of the categories of things that exist or may exist in some domain. The product of such a study is a catalog of the types of things that are assumed to exist in a domain of interest D from the perspective of a person who uses a language L for the purpose of talking about D;

2.2.11.   Noy and McGuinness (2003): an ontology is a formal explicit description of concepts in a domain of discourse, properties of each concept describing various features and attributes of the concept, and restriction on slots.

2.3   Clarification of Terms

Some clarifications are needed in order to get the proper meanings of the definitions above.

"Conceptualizations" are a "set of elements, considered as existing in some, together with a set of concepts [such as categories that divide up the reality into objects, processes, and classes] and relationships which characterize (or enable to understand or to describe) that portion of reality, from a particular perspective" (Goy and Magro 2015). According to Guarino and Giaretta (1995), a conceptualization is basically the idea of the world that a person or a group of people can have. It is given by a set of rules (formally) constraining the structure of a piece of reality, in order to organize relevant objects and relations. Finally, it can thus be described as independent from the used vocabulary and from the occurrences of a specific situation (Guarino and Giaretta 1995).

"Explicit" means that the type of concepts (and terms) used, and the constraints on that use are explicitly defined (in a generic and formal way).[5]

"Formal" refers to the fact that the ontology should be machine-readable (suitable for automated reasoning) and, if not directly human-readable, it should at least contain plain text notices or explanations of concepts and relations for the human user.[6]

"Shared" reflects the notion that an ontology captures consensual knowledge, that is, it is not private for some individuals but, at least to some extent, accepted by some community of people, even though not universally.[7]

"Partial account" indicates that we represent the domain of interests with a certain perspective: an ontology entails some sort of world view with respect to the given domain (Guarino and Giaretta 1995; Guarino 1998). This domain can be as specific

---

[5] See Uschold (1996), Fensel (2001), Zelewski et al. (2001), Mizoguchi (2003).

[6] See Borst (1997), Guarino (1998), Studer et al. (1998), Uschold (1998), Fikes et al. (1999), Sowa (2000).

[7] See Studer et al. (1998).

as a single task or application, always remaining "some part of a conceptualization" (Uschold 1996).

"Specification" points to the fact that an ontology is expressed by an intensional semantic structure (i.e., a *logical theory*), which entails some sort of world view,[8] or by means of a (logical) language (e.g., in a language belonging to the first-order predicate calculus and suitable for a practical implementation of reasoning mechanisms) which contributes a reduction of ambiguity in the knowledge representations.

This logical theory is composed of a "vocabulary" (human-understandable definitions of the objects in natural language) used to describe the reality at hand, and a set of explicit assumptions or axioms (Mizoguchi 2003). Typically, the vocabulary (the *modeling primitives*) of an ontology is contained in a taxonomy which already includes classes, simple relations, and axioms (Bullinger 2008, p. 139).[9]

2.4 From the Ontological Turn to Taxonomies of Philosophical Ontologies

As we have seen in Chap. 1, IT/computer ontology represents only one of the main areas of research that make up the ontological background of applied ontology of geography, which also includes philosophy and geography.

From a philosophical point of view, the "ontological turn" describes the recent and progressive proliferation of ontological research, specifically in the analytic area.[10] The development of this research has been characterized by an increasing number of hypotheses and methods of investigation, from which a heterogeneous debate has followed. The consequent plurality of guidelines has made it difficult to provide an exhaustive classification of the various and different positions at stake.

D'Agostini (2002), for example, distinguishes among conceptual, modal, and naturalistic ontologies. Conceptual ontologies start from the question of the contextuality of language[11]; modal ones are connected to the theory of possible worlds[12]; naturalistic ontologies are linked to the thesis of Quine (1981) and to Australian Realism,[13] which in turn are intertwined with empirical sciences and, in particular, with physics.

Otherwise, the taxonomy of Runggaldier and Kanzian (1998) includes modal and naturalistic ontologies but classifies Strawson's position among descriptive ontologies that comprehend the proposals of all of those contemporary authors oriented

---

[8] See Guarino and Giaretta (1995), Uschold and Grueninger (1996), Guarino (1998), Goy and Magro (2015).

[9] See Zelewski et al. (2001), Hesse (2002), Mizoguchi (2003), Krcmar (2005).

[10] See in particular Martin and Heil (1999).

[11] See Strawson (1959), Jackson (1998).

[12] See Lewis (1986).

[13] See Armstrong (1989, 1997).

toward Aristotle's metaphysics.[14] Moreover, the authors introduce the phenomeno-
logical ontologies, which are developed from the perspectives of Brentano and
Husserl. Among them, Runggaldier and Kanzian further distinguish among:

2.4.1.   mereologists,[15] which are connected to the Polish logic;
2.4.2.   authors who emphasize the fundamental role of intentionality, philosophy
         of language, and theory of knowledge[16];
2.4.3.   anti-reductionists, who do not think that all the entities may belong to only
         one all-embracing category and suggest a division into a collections of
         categories.[17]

   Another fundamental division in the realm of analytical ontology is that between
formal and material ontologies. Such a division draws quite different areas of research
characterized by different tools and conceptual background. Formal ontology is
concerned with the task of laying bare the formal structure of all there is, whatever it is. No matter what our domain of quantification includes, it must exhibit some
general features and obey some general laws, and the task of the ontology would be
to figure out such features and laws. More generally, it should pertain to the aim of
formal ontology to work out a general theory of formal relations such as identity,
parthood and dependence (Varzi 2011a, pp. 3–4). Conversely, material ontology deals
with the question of what there is and is aimed at drawing up a detailed and exhaustive
inventory of what exists. Moreover, material ontology is taken to be closely related to
the (specific) aspects of reality studied by different scientific and social disciplines.
In particular, scientific theories and their results (as well as the analysis of linguistic
structures) constitute its fundamental conceptual tools and background (Varzi 2005,
p. 33).

2.5 Ontology, Metaphysics, and Scientific Disciplines

This plurality of positions does not prevent having a generally shared meaning of
the term "ontology", at least in the analytic area. Ontology denotes the philosophical
discipline concerned with the question of *what entities exist*—a task that is often
identified with that of drafting a complete and detailed inventory of the universe.
In this way, ontology (the big-O Ontology of Chap. 1) is depicted as the science of
being, that is the discipline that, by using logical and empirical methods, focuses
on the totality of (kinds of) entities "which make up the world on different levels
of focus and granularity, and whose different parts and aspects are studied by the
different folk and scientific disciplines" (Smith 2004, p. 158).

---

[14] See Körner (1974, 1984), Lowe (2006).
[15] See Mulligan (2000), Simons (1987, 1994), Smith (1997).
[16] See Chisholm (1976, 1984, 1996).
[17] See Bergmann (1967), Grossmann (1992), Tegtmeier (1992).

In the analytic area, it is also common to think of ontology as a proper part of metaphysics (that part that has to do with what there is) and to consider ontology in some way prior over metaphysics. Ontology aims at establishing "what there is", whereas metaphysics is the study of "what it is": that is, the discipline which seeks to explain the ultimate nature (and the necessary features) of the items included in the inventory of what there is, and the reasons why there is what there is. More precisely, one must first of all figure out what (kinds of) things exist (or might exist); "then one can attend to the further question of what they are, specify their nature, speculate on those features that make each thing the thing it is" (Varzi 2011b, p. 408). Obviously, the close interdependence between these two disciplines makes a partition of the respective goals very difficult: it is not clear how to establish what there is without saying what it is.[18] Nowadays, there is much debate on the disciplinary distinction between ontology and metaphysics, however there is no high agreement about how to draw such a distinction.[19]

But, by dealing with folk and scientific disciplines, ontology can also be viewed as an investigation on the ontological commitments or premises embodied in different scientific theories as well as in common sense and as an analysis of the categorial and hierarchical structure of reality. The latter aspect specifically regards the basic constituents of reality (entities such as objects, properties, relations, events, processes, etc.), and the structural relationships among them. On this basis, the connection between ontology and (the results of different) scientific disciplines has been implemented along two directions. On the one side, the improvement has allowed us to make explicit the ontological assumptions and commitments of these non-philosophical disciplines. On the other side, it has led to a proliferation of regional ontologies,[20] aimed at providing an inventory of what there is within the domain of each specific discipline. The non-reductionist hypothesis embraced by regional ontologies is that the (fundamental) entities postulated by different disciplines are irreducible to the entities postulated by other disciplines, deserving a specific and separate study, so to increase our explanatory resources.

A critical point arises from the relationship between common sense and scientific theories. Specifically, how can we define such a relationship? Is there a connection between common sense and scientific theories or maybe is there a gap? Must the ontology define what there is according to common sense, scientific theories or either of them? May we use the same ontological categories for describing both? Or again, must we have different categories in order to portray these distinct and separate levels of reality? In answering these questions, it should be emphasized that, in the ontological debate, the two horns of this relationship have been named differently, while expressing quite similar meanings. Sellars (1963), for example, distinguishes between manifest and scientific images; Smith and Mark (2001) divide between folk and scientific domains; Cumpa (2014) uses the expressions ordinary world and physical universe. By "ordinary word", he understands "an ordinary level

---

[18]See Ferraris (2008, pp. 16–7); Bianchi and Bottani (2003).
[19]See Berto (2010).
[20]See Ferraris (2008).

of thinghood with which ordinary people are acquainted in their commonsensical and practical experiences". By "physical universe", he means "a scientific level of thinghood with which scientists are acquainted in their experimental research, such as fundamental physics, chemistry, or biology" (Cumpa 2014, pp. 319–20). Then, by using the terminology of Cumpa, we can distinguish two main positions about the relation between these two horns. The first one maintains that the ordinary world and the physical universe are levels of thinghood *isolated* from each other, and so embraces a non-reductionist perspective for the categories we need for describing these separate levels. The second position, in contrast, thinks of the world as a whole composed of the ordinary world *and* the physical universe. And thus, it requires some ontological categories that have the explanatory power to account for the world as a complex composed of ordinary entities and entities of scientific disciplines, and for the relation between the ordinary world and the physical universe.

2.6 Ontology of Geography

Among regional ontologies, the ontology of geography owes its development primarily (but not exclusively) to the pioneering works of Casati, Mark, Smith, and Varzi. According to Smith and Mark (2001), this kind of ontology might be defined as that part of philosophical ontology which analyzes the mesoscopic world of geographical partitions in order to:

2.6.1.   establish whether and what kinds of geographical entities exist, and how they can be defined and classified in an ontological system that gathers them together;
2.6.2.   develop a theory of spatial representation;
2.6.3.   explaining whether and how the geographic descriptions of reality emerging from common sense (the small-g geography of Chap. 1) can be combined with those derived from academic geography (the big-G Geography of Chap. 1).

According to 2.6.1 and 2.6.2, the ontology of geography is also defined as that discipline that specifically studies: geographic entities (entities such as mountains, oceans, countries, etc.), their borders (natural and/or artificial, regardless of the fact that these boundaries might be part of the entities they define), their spatial representation (in maps, software, etc.), their mereological and topological relations, and their location.

Regarding 2.6.3, Smith and Mark also specify that:

Mesoscopic geography deals mostly with qualitative phenomena, with phenomena which can be expressed in the qualitative terms of natural language; the corresponding scientific disciplines, in contrast, deal with the same domain but consider features which are quantitative and measurable. GIS thus requires methods that will allow the transformation of quantitative geospatial data into the sorts of qualitative representations of geospatial phenomena that are tractable to non-expert users—and for this […] we need a sound theory of the ontology of geospatial common sense. [...] One of the most important characteristics of the geographical domain is the way in which geographical objects are not merely located in space, but are typically parts of the Earth's surface, and inherit mereological properties from that surface. At the same time, however, empirical evidence suggests that geographical objects are organized into categories in much the same way as are detached, manipulable objects (Smith and Mark 2001, p. 596).

2.7 Common-Sense Geography

As 2.6.3 refers to geographic descriptions of reality shaped on common sense, we should now specify what is meant precisely by such descriptions. In this regard, it is worth mentioning Smith and Mark's experiments that show how non-expert subjects conceptualize geospatial phenomena and exchange geographical information.

Before summarizing procedures and results of these experiments, it is, however, necessary to step back and examine Smith and Mark's theoretical assumptions, which are essentially two:

2.7.1. the fundamental role of folk disciplines, or common-sense conceptualizations;

2.7.2. the organization of these conceptualizations in terms of systems of objects falling under categories.

Regarding 2.7.1, we should emphasize that, in general, we can associate *pre-scientific* or *folk* counterparts with many scientific disciplines. The subject-matter of these counterparts is the common-sense reality and the (tacit) conceptualizations behind it:

> the study of folk conceptualizations [...] may also be of interest in helping us to provide better theories of common-sense reasoning, for if common-sense reasoning takes place against a background of common-sense beliefs and theories, then we cannot understand the former unless we also develop good theories of the latter. The study of the ways non-experts conceptualize given domains of reality might then help us also in our efforts to maximize the usability of corresponding information systems [...]. Geographical concepts shared in common by non-experts represent a good conceptualization [...]. They are transparent to reality: mountains, lakes, islands, roads truly do exist, and they have the properties we commonsensically suppose them to have. The task of eliciting this folk ontology of the geographic domain will turn out to be by no means trivial, but we believe that the effort invested in focusing on good conceptualizations in the geographical domain will bring the advantage that it is more likely to render the results of work in geospatial ontology compatible with the results of ontological investigations of neighbouring domains. It will have advantages also in more immediate ways, above all in yielding robust and tractable standardizations of geographical terms and concepts (Smith and Mark 2001, p. 595).

As a result, common-sense geography (CSG) became a topic of discussion in the '90 s when software developers tried to design virtual spaces according to objective parameters that differ from human perception and experience. But, what do we intend when we speak of CSG?

Geus and Thiering (2014) sketch some of its features, which might be resumed as follows. CSG:

2.7.3   denotes the ways non-experts conceptualize geography, in terms of beliefs, theories, and knowledge;

2.7.4   concerns the belief about general regularities in the mesoscopic domain and the consensus of an epistemic collective or community—so, it is to be understood as *shared* knowledge and beliefs;

2.7.5   refers to a *naïve* perception and description of space and the use of *intuitive* arguments in geographical contexts;

2.7.6   is transparent to reality and accessible also for non-expert users;

2.7.7   has been and for the most part still is dismissed at best as a sort of pre- or sub-scientific *knowledge*;

2.7.8   denotes a *lower* geography, to be distinguished from *professional* or *higher* geography, that is, the phenomenon of the spread and application of geographical knowledge outside of expert circles and disciplinary contexts;

2.7.9   consists of naïve physics and folk psychology, and it is strictly related to (physical-geographic) mesoscopic phenomena that is quite independent from our knowledge and culture, and immediately accessible to human beings in everyday perception and actions.

2.8   Primary and Secondary Theories

2.7.9 is crucial also for Smith and Mark's experiments. Indeed, by following Horton's (1982) distinction between primary and secondary theories (and beliefs), the authors focus only on those specific common-sense conceptualizations that Horton includes among primary theories.

Primary theory is that part of common sense that we can find in all cultures and in all human beings at all stages of their development. It just consists of naïve physics and basic folk psychology: that is, the total stock of basic theoretical beliefs which all humans need in order to perceive and act in everyday contexts. And such a common sense relates to mesoscopic phenomena in the realm that is immediately accessible to perception and action. Secondary theory, in contrast, concerns those collections of folk beliefs that are characteristic of different cultures, communities, and economic settings. Moreover, according to Horton, secondary theory relates to phenomena that are either too large or too small to be immediately accessible to human beings in everyday perception and actions or to objects and processes that are, otherwise, hidden.

Agreement in primary theory has evolutionary roots: there is a sense in which the theory about basic features must correspond to the reality which it purports to represent, for if it did not do so, its users down the ages could scarcely have survived. At the same time, its structure has a fairly obvious functional relationship to specific human aims and to the specific human equipment available for achieving these aims. […] The commonsensical world as the world that is apprehended in primary theory is thus to a large degree universal. It is apprehended in all cultures as embracing a plurality of enduring substances possessing sensible qualities and undergoing changes (events and processes) of various regular sorts, all existing independently of our knowledge and awareness and all such as to constitute a single whole that is extended in space and time. This body of belief about general regularities in the mesoscopic domain is put to the test of constant use, and survives and flourishes in very many different environments. Thus no matter what sorts of changes might occur in their surroundings, human beings seem to have the ability to carve out for themselves, immediately and spontaneously, a haven of commonsensical reality. Moreover, our common-sense beliefs are readily translated from language to language, and judgments expressing such beliefs are marked by a widespread unforced agreement (Smith and Mark 2001, p. 597).

Primary theory is generally organized in terms of categorical systems of objects falling under categories, typically determined by prototypical instances. Usually, these systems are organized hierarchically in the form of a tree (they have only one all-embracing category), "with more general categories at the top and successively more specific categories appearing as we move down each of the various branches" (Smith and Mark 2001, p. 601). Deviations from the tree structure are occasionally proposed, for example, systems which do not have one all-embracing category but a collection of trees (a forest).

The question has still not been resolved whether the structure and organizing principles governing our cognitive categories remain the same as we move beyond these families of examples to objects at geographical scales. This holds, too, in regard to the primary theory of phenomena in the geographic domain, which is organized around categories such as mountain, lake, island. The primary axis of a folk ontology is its system of objects. This holds, too, in the realm of geospatial folk categories. The attributes (properties, aspects, features) and relations within the relevant domain form a secondary axis of the ontology, as also do events, processes, actions, states, forces and the like. The system of objects remains primary, however, because attributes are always attributes of objects, relations always relations between objects, events always events involving objects, and so forth, in ways which imply a dependence of entities in these latter categories upon their hosts or bearers in the primary category of objects (Smith and Mark 2001, p. 601).

Finally, the basic categories of primary theory are identified on empirical and cognitive grounds, play a special role in common-sense reasoning and represent a theoretical compromise between two different aims: cognitive economy and informativeness—regarding the latter point, the notions of explanation and causality play a fundamental role.

2.9   Mark and Smith's Experiments

Having highlighted the importance of common-sense geographical conceptualizations, organized hierarchically in categorical systems with the form of a tree, it is decisive to clarify what (kinds of) geographical entities might populate such systems. To this end, Smith and Mark's experiments can be specifically incisive, as they are designed to establish how non-expert subjects conceptualize geospatial entities in response to a series of differently phrased geographical elicitations.

Without dwelling on the scientific debt to Battig and Montague's experiments (1968), it could be enough to point out here the common procedural approach in the method of elicitation embraced in both studies: subjects were given a series of categorical names and were then asked to write a list of entities that might belong to those categories, in whatever order they came to mind. To be more precise,

> subjects were tested simultaneously in a large classroom, at the beginning or end of a lecture. Subjects were students in two large sections of a first-year university course called "World Civilization". Versions 1 to 5 of the experiment were administered in one class- room, and versions 6 to 10 in the other. Versions 6–10 differed from versions 1–5 only in the order of presentation of stimuli, so that we could test for intercategory priming effects. Within each class, the five versions were printed on different colours of paper, and handed out from piles interleaving the five versions, in order to maximize the chance that the subject pools for the five versions were as similar as possible. Subjects were given a series of nine category names, each printed at the top of an otherwise blank page. They were asked to wait before turning to the first category, and then to write as many items included in that category as they could in 30-seconds, in whatever order the items happened to occur to them. After each 30-second period, they were told to stop, turn the page, and start the next category. A total of 263 subjects completed the first geographic category, with between 51 and 56 subjects responding to each version of the survey. Chi-squared tests showed no significant differences between responses from the two classrooms for any of the questions. Following a pre-test reported by Mark et al. (1999), we chose nine categories to test with larger numbers of subjects. The first category tested was a non-geographic category (a chemical), which we hoped would provide a neutral, unprimed basis for the remaining questions. This was followed by a somewhat neutral phrase, a type of human dwelling. The third stimulus give to the subjects presented one of five variations on the phrase a kind of geographic feature. [...] We therefore formulated five different wordings of our target phrase, and presented these alternative wordings to five different groups of subjects, in effect changing the base noun of the superordinate category (Smith and Mark 2001, pp. 604–5).

According to Smith and Mark, instances with higher consensus ("norms") would provide indicators ("prototypical example") for basic geographical categories, or even a first approximation of the basic noun lexicon for geographical ontologies. Moreover,

> higher consensus among the lists of examples from one subject to another is an indicator that the category in question is a natural category in the sense of a category that is rooted more firmly in our cognitive architecture than are categories offered for elicitation which produce lower consensus or no consensus at all (Smith and Mark 2001, p. 603).

2.10   Analysis of Results

Figure 2.1 summarizes the results of the experiments, reporting tests carried out in the USA at the University of Buffalo and targeted exclusively to native American English speakers. Thereafter, the experiments have been repeated in Finland, Croatia, and in the UK, and have produced similar trends even among non-native American English speakers suggesting that the outcomes should not be considered as an artifact of a particular pool of subjects or of a specific language (Smith and Mark 2001, p. 610).

In Fig. 2.1, we show, in bold, the geographical categories used for the elicitation and the number of subjects involved in the experiments. Under each category, we specify the (respective) geographical entities mentioned with a statistically more significant frequency and the number of elicitations received.

a kind of geographic feature (54)	a kind of geographic object (56)	a geographic concept (51)	something geographic (51)	something that could be portrayed on a map (51)
Mountain (48)	Mountain (23)	Mountain (23)	Mountain (32)	River (31)
River (35)	River (18)	River (19)	River (26)	City (30)
Lake (33)	Map (17)	Ocean (16)	Lake (25)	Road (27)
Ocean (27)	Ocean (16)	Sea (11)	Ocean (18)	Mountain (25)
Valley (21)	Lake (13)	Lake (10)	Hill (11)	Country (23)
Hill (20)	Globe (11)	Continent (9)	Map (11)	Lake (21)
Plain (19)	Continent (10)	Plateau (8)	Sea (9)	Ocean (18)
Plateau (17)	Peninsula (10)	Map (7)	Continent (8)	State (15)
Desert (14)	Hill (9)	Valley (7)	Country (8)	Continent (12)
Volcano (10)	Compass (8)	Island (7)	Island (7)	Street (8)
Sea (9)	Sea (8)	Delta (6)	Land (6)	Town (8)
Peninsula (8)	Valley (7)		Plateau (6)	Highway (7)
Island (8)	Island (7)		Desert (6)	Park (6)
	Rock (6)		The world (5)	
	Atlas (6)			

Fig. 2.1 Results of experiments

In summarizing the results of the experiments, it should be pointed out that geographical-physical entities such as mountains, rivers, and oceans show much higher consensus than artifacts (or *fiat* entities) derived from social sciences, cultural studies and economics.

> Only five terms reached the 10% threshold on all five versions of this question and all of these are physical: mountain, river, lake, ocean, and sea. This suggests that, for this population of subjects at least, it is the physical environment that provides the most basic examples of geographical phenomena. This predominance of physical geography lends support to the view that concepts for (some) types of geographical objects are very deeply rooted in our primary-theoretic cognitive architecture, namely those—like mountain, river, lake, ocean, and sea—referring to objects of a kind which were (surely) strongly relevant to the survival of our predecessors in primeval environments (Smith and Mark 2001, p. 606).

And yet, the experiments also show significantly different responses, according to the categories used for their elicitation. Moreover, of all the five categories, the responses to "a kind of geographic feature" still stand out as most strongly dominated by aspects of the physical environment. Conversely, entities derived from the domain of human geography (city, road, country, state) appear far more often in response to "something that could be portrayed on a map" than to any other category.[21] "Geographic object" stands out from the other categories listed in the degree to which it elicits examples of *small, portable* items, such as map, globe, compass, or atlas, that are listed more frequently under this than under any other heading. The entities elicited by the category "a geographic concept" manifest the lowest degree of internal coherence of all the five categories. Finally, "something geographic"

---

[21] In regard to this category, the authors emphasized that geographers "are not studying geographical things as such things are conceptualized by naïve subjects. Rather, they are studying the domain of what can be portrayed on maps" (Smith and Mark 2001, p. 609).

picks up a mixture of the responses typical of the other categories—in particular of "geographic feature". However, entities such as map, the world and land (albeit with a markedly lower frequency) are more frequent here than for any of the other categories.

In conclusion, the authors point out that the results of the experiments have proved that the terms reported by the subjects basically denote geographical entities, which can provide a first approximation of the basic noun lexicon for describing the reality emerging from GSG. Still, some difficulties arise in making explicit the relationship between ontology and language. In particular, Smith and Mark have emphasized that the terms can be intersected in different ways depending on the categories used for their elicitation. Problems can thus be found in extracting a single hierarchy of classes for the specified terms and in determining which term should be used for the superordinate category within each geographical entity to be included.

> Thus some of the terms suggested can be held to narrow the scope of ontology illegitimately to some one particular *kind* of being, for example to beings which *exist*, or are *real*, or come ready-demarcated into *items*. Similar arguments have also been seen in the international spatial data standards community. Given the particular meanings of the terms object, entity, and feature in the US Spatial Data Transfer Standard, for example, how should these terms be translated into other natural languages? […] We suggest that these conceptualizations represent not different *ontologies* that we might ascribe to the subjects in the groups we tested. Rather, they are a matter of different superordinate categories— objects, features, things— that intersect to varying degrees in virtue of the fact that they share a common domain—the domain of geography. Particular kinds of phenomena, such as mountains or maps or buildings, have different relative prominence or salience under these different superordinate categories. We propose, therefore, that there is just one (folk) ontology of the geospatial realm, but that this ontology gets pulled in different directions by contextually determined salience conditions. […] This outcome is significant not least because the distinctions captured by ontological terms are commonly held to be of low or zero practical significance (Smith and Mark 2001, pp. 610–1).

References

Armstrong DM (1989) Universals: an opinionated introduction. Westview Press, Boulder

Armstrong DM (1997) A world of states of affairs. Cambridge University Press, Cambridge

Battig WF, Montague WE (1968) Category norms for verbal items in 56 categories: a replication and extension of the Connecticut norms. J Exp Psychol Monograph 80(2):1–46

Bergmann G (1967) Realism XE "Realism" . University of Wisconsin Press, Madison

Bernaras A, Laresgoiti I, Corera J (1996) Building and reusing ontologies for electrical network applications. In: Proceedings of the European conference on artificial intelligence (ECAI_96), pp 298–302. Budapest, Hungary

Berto F (2010) L'esistenza non è logica. Laterza, Roma, Bari

Bianchi C, Bottani A (eds) (2003) Significato e ontologia. Franco Angeli, Milano

Borst WN (1997) Construction of engineering ontologies. Centre for telematica and information technology. University of Tweenty, Enschede, The Netherlands

Bullinger A (2008) Innovation and ontologies. Structuring the early stages of innovation management. Gabler, Wiesbaden

Chisholm RM (1976) Person and object. Open Court, La Salle

Chisholm RM (1984) The primacy of the intentional. Synthese 61:89–109

Chisholm RM (1996) A realistic theory of categories. Cambridge University Press, Cambridge

Cumpa J (2014) A materialist criterion of fundamentality. Am Philos Q 51(4):319–324

D'Agostini F (2002) Che cosa è la filosofia analitica? In: D'Agostini F, Vassallo N (eds) Storia della filosofia analitica. Einaudi, Torino, pp 3–76

Fensel D (2001) Ontologies: a silver bullet for knowledge management and electronic commerce. Springer, Berlin, Heidelberg, New York

Ferraris M (ed) (2008) Storia dell'ontologia. Milano, Bompiani

Fikes R, Farquhar A (1999) Distributed repositories of highly expressive reusable ontologies. IEEE Intell Syst 14(2):73–79

Geus K, Thiering M (2014) Common sense geography and mental modelling: Setting the stage. In: Geus K, Thiering M (eds) Features of common sense geography. Implicit knowledge structures in ancient geographical texts. LIT, Wien

Gómez-Pérez A, Fernández-López M, Corcho O (2004) Ontological engineering: with examples from the areas of knowledge management, E-Commerce and the semantic web. Springer, Berlin, Heidelberg, New York

Goy A, Magro D (2015) What are ontologies useful for? Encyclopedia of information science and technology. IGI Global, pp 7456–7464

Grossmann R (1992) Existence of the world: an introduction to ontology. Routledge, London, New York

Gruber TR (1993) A translation approach to portable ontology specifications. Knowl Acquis 5(2):199–220

Guarino N (1998) Formal ontology and information systems. In: Proceedings of FOIS '98. Trento, Italy. IOS Press, Amsterdam, pp 3–15

Guarino N, Giaretta P (1995) Ontologies and knowledge bases—towards a terminological clarification. In: Mars NJ (ed) Towards very large knowledge bases—knowledge building and knowledge sharing. IOS Press, Amsterdam, pp 25–32

Hesse W (2002) Ontologie(n). Informatik Spektrum 25(6):477–480

Horton R (1982) Tradition and modernity revisited. In: Hollis M, Luke S (eds) Rationality and relativism. Blackwell, Oxford, pp 201–260

Jackson G (1998) From metaphysics to ethics: a defence of conceptual analysis. Oxford University Press, Oxford

Jaziri W, Gargouri F (2010) Ontology theory, management and design: An overview and future directions. In: Gargouri F, Jaziri W (eds) Ontology theory, management and design: advanced tools and models. Information Science Reference, Hershey

Krcmar H (2005) Informations management. Springer, Berlin, Heidelberg, New York

Körner S (ed) (1974) Practical reason. Yale University Press, New Haven

Körner S (1984) Metaphysics: its structure and function. Cambridge University Press, Cambridge

Lewis D (1986) On the plurality of worlds. Blackwell, Oxford

Lowe EJ (2006) The four-category ontology: a metaphysical foundation for natural science. Clarendon Press, Oxford

Mark DM, Smith B, Tversky B (1999) Ontology and geographic objects: An empirical study of cognitive categorization. In: Freksa C, Mark DM (eds) Proceedings of spatial information theory. Cognitive and computational foundations of geographic information science. International conference COSIT'99 Stade, Germany. Springer, Berlin, Heidelberg, pp 283–298

Martin CB, Heil J (1999) The ontological turn. In: French PA, Wettstein HK (eds) New directions in philosophy, midwest studies in philosophy, vol 23, pp 34–60

Mizoguchi R (2003) Tutorial on ontological engineering: part 01: introduction to ontological engineering. N Gener Comput 21(4):365–384

Mulligan K (2000) Ontologie e Métaphysics. In: Engel P (ed) Précis de Philosophie analytique. Puf, Paris

Neches R, Fikes RE, Finin T, Gruber TR, Senator T, Swartout WR (1991) Enabling technology for knowledge sharing. AI Mag 12(3):36–56

Noy NF, McGuinness DL (2003) Ontology development 101: a guide to creating your first ontology. Stanford University, Stanford (CA)

Quine WVO (1981) Theories and things. Harvard University Press, Cambridge

Runggaldier E, Kanzian C (1998) Grundprobleme der analytischen ontologie. Verlag Ferdinand Schöning, Paderborn

Sellars W (1963) Philosophy and the scientific image of man. In: Sellars W (ed) Science, perception, and reality. London, Routledge, Kegan Paul, pp 1–40

Simons P (1987) Parts: an essay in ontology. Clarendon Press, Oxford

Simons P (1994) Particulars in particular clothing: three trope theories of substance. Res 54:553–575

Smith B (1997) On substance, accidents and universals: in defence of a constituent ontology. Philos Pap 27:105–127

Smith B (2004) Ontology. In: Floridi L (ed) The Blackwell guide to the philosophy of computing and information. Blackwell, Malden (MA), pp 155–166

Smith B, Mark DM (2001) Geographical categories: an ontological investigation. Int J Geogr Inf Sci 15(7):591–612

Sowa JF (2000) Guided tour of ontology. http://www.jfsowa.com/ontology/guided.htm

Strawson P (1959) Individuals. An essay in descriptive metaphysics. Methuen, London

Studer R, Benjamins VR, Fensel D (1998) Knowledge engineering: principles and methods. IEEE Trans Data Knowl Eng 25(1–2):161–197

Swartout B, Ramesh P, Knight K, Russ T (1997) Toward distributed use of large-scale ontologies. In: AAAI symposium on ontological engineering. Stanford, CA

Tegtmeier E (1992) Grundzüge einer kategorialen Ontologie: Dinge, Eigenschaften, Beziehungen, Sachverhalte. Alber, Freiburg, München

Uschold M (1996) Building ontologies: towards a unified methodology. Technical Report of the Artificial Intelligence Applications Institute, No. 197. Edinburgh, Scotland

Uschold M (1998) Knowledge level modelling: concepts and terminology. Knowl Eng Rev 13(1):5–29

Uschold M, Grueninger M (1996) Ontologies: principles, methods and applications. Technical Report of the Artificial Intelligence Applications Institute, No. 191. Scotland, Edinburgh

Uschold M, Jasper R (1999) A framework for understanding and classifying ontology applications. In: Proceedings of the IJCAI99 workshop on ontologies and problem-solving method. Stockholm, Sweden

Varzi AC (2005) Ontologia. Laterza, Rome, Bari

Varzi AC (2011a) On the boundary between material and formal ontology. In: Smith B, Mizoguchi R, Nakagawa S (eds) Interdisciplinary ontology, vol 3, pp 3–8. Keio University, Tokyo

Varzi AC (2011b) On doing ontology without metaphysics. Philos Perspect 25:407–423

Zelewski S, Schuette R, Siedentopf J (2001) Ontologien zur Repraesentation von Domaenen. In Schreyoegg G (ed) Wissen in Unternehmen. Konzepte, Massnahmen, Methoden. Erich Schmidt Verlag, Berlin, pp 183–221

Part II
Systematizing the Geographical World

Chapter 3
Spatial Representation

Abstract Chapter 2 argued that the main goals of ontology of geography includes developing a formal theory of spatial representation, with special reference to geographical phenomena. This chapter undertakes to offer an introduction to the theoretical tools needed for advancing such a formal theory. First, we show that the list of tools can include mereology and topology as well as the geo-ontological distinction between classical and non-classical geographies. Second, following Casati et al. (1998), we argue that classical geography describes a robust way of *tiling* regions in the presence of three general axioms: (1) Every single geographic entity is located at some unique spatial region. (2) Every spatial region has a unique geographic entity located at it. (3) If two entities are located are at the same spatial region, then they are the same entity. Third, we maintain that any geography can be considered as non-classical if it drops and/or adds axioms to those of classical geography. Finally, we showcase some possible issues emerging from the application of the distinction between classical and non-classical geographies to the cartographic representation.

Keywords Cartographic representation · Classical geography · Geographical entities · Mereology · Non-classical geographies · Ontology of geography · Spatial representation · Topology

3.1 Ontology of Geography and Spatial Representation

Chapter 2 defined ontology of geography as that part of the philosophical ontology concerned with the mesoscopic world of geographical partitions and aimed at the following:

3.1.1. explaining how the geographic descriptions of reality emerging from common sense can be combined with those derived from academic geography (Geus and Thiering 2014);

© The Author(s), under exclusive license to Springer Nature Switzerland AG 2021   39
T. Tambassi, *The Philosophy of Geo-Ontologies*, SpringerBriefs in Geography,
https://doi.org/10.1007/978-3-030-78145-3_3

3.1.2. establishing what kinds of geographical entities exist, and how they can be
defined and classified in an ontological system that gathers them together
(Smith and Mark 2001);

3.1.3. developing a formal theory of spatial representation, with special reference
to spatial phenomena on the geographic scale (Casati et al. 1998).

Here, I intend to maintain that 3.1.2 and 3.1.3 are strictly connected: in particular,
3.1.3 may be thought of as dependent upon 3.1.2. Indeed, as Casati and Varzi (1999)
remark, the advance of a theory of spatial representation should be combined with (if
not grounded on) an account of the kinds of (geographical) entity that can be located
or take place in space. In short, this would mean providing a definition of what may
be collected under the rubric of *spatial entities* but also outlining how to distinguish
them from purely *spatial items* (such as points, lines, regions, and so forth).

> What special features make them spatial entities? How are they related to one another, and
> exactly what is their relation to space? On the methodological side, the issue is the definition
> of the basic conceptual *tools* required by a theory of spatial representation, understood as
> a theory of the representation of these entities. There may be some ambiguity here, due
> to a certain ambiguity of the term "representation". We may think of (1) a theory of the
> way a cognitive system represents its spatial environment (this representation serving the
> twofold purpose of organizing perceptual inputs and synthesizing behavioral outputs), or
> (2) a theory of the spatial structure of the environment […]. The two notions are clearly
> distinct. Presumably, one can go a long way in the development of a cognitive theory of type
> 1 without developing a formal theory of type 2, and *vice versa*. However, both notions share
> a common concern; both types of theory require an account of the geometric representation
> of our spatial competence before we can even start looking at the mechanisms underlying
> our actual performances. (Casati and Varzi 1999, pp. 1–2)

Moreover, developing such a formal theory also implies choosing between absolutist and relational theories of space. The former maintains that the space exists as
an independently subsistent individual (a sort of container) over and above its inhabitants (objects, events and spatial relations between objects and events, or without
all these entities). Conversely, the relational theory considers that spatial entities are
cognitively and metaphysically prior to space. Thus, there is no way to identify a
region of space except by reference to what is or could be located at that region
(Casati and Varzi 1999, p. 1).

3.2 Tools for Spatial Representation

In order to enhance a theory of spatial representation, the ontology of geography
has developed three main theoretical tools strictly interconnected and mutually
interacting:

3.2.1. mereology;

3.2.2. topology;

3.2.3. theory of spatial location.

Mereology, in general, is the theory of parthood relations (Simons 1987; Smith and Mark 1998; Casati and Varzi 1999; Mark et al. 1999). It also includes some temporal parameters, which help us to specify the criteria of identity for the geographical entities and for their constitutive parts. Topology is the theory of qualitative spatial relations, such as continuity, contiguity, connection, overlapping, containment, distance, separation, and so forth (Smith 1994, 1995, 1996; Smith and Varzi 2000; Varzi 2007). Therefore, it examines notions like boundary and border, their spatial and temporal relations and their relationships with the entities they connect and circumscribe—in this sense, topology is also strictly related to geometry and morphology.[1] Finally,the theory of spatial location deals with the relationship between an entity and the spatial region (of space) it occupies or in which it is located. In a strict geographical sense, this relation is not the one of identity—a geographical entity is not identical with the spatial region it occupies, besides two or more different geographical entities can share the same location at the same time—and it does not imply that any single geographical entity is located somewhere, or that any spatial region is the location of a geographical entity (Casati et al. 1998; Varzi 2007).

3.3 Classical and Non-Classical Geographies

In addition to these tools, Casati et al. (1998) also introduce the distinction between classical and non-classical geographies that, according to the authors, can be useful for the specification of the (kind of) geography behind the spatial representations.

Starting from the assumption that there is no single universally recognized formulation that precisely indicates what classical geography is,[2] the authors characterize a geography on a region R as a way of assigning (via the location relation) geographic objects of given types to parts or sub-regions of R. Then, they propose to put forward some principles for a minimal characterization of geographic representation. These principles are such that the violation of one of those principles produces intuitively incomplete representations.

Under these assumptions, Casati and colleagues sustain that their idea of classical geography (CG) does not carry any normative claim. It simply describes a rather robust way of *tiling* regions in the presence of certain general axioms, which specify that:

---

[1]For an analysis of the connection between mereology and topology and of the notion of mereotopology, see Smith (1995), Breysse and De Glas (2007). For an analysis of the relation between the notions of topology and border, see Casati et al. (1998), Smith and Varzi (2000), Varzi (2007).

[2]Among the most significant works that investigate the notion of "classical geography" in a strict geographical sense and analyze its relations with the concepts of spatial location and representation, see Lukermann (1961), Geus and Thiering (2014).

CG1.   every single geographic entity (nations, lakes, rivers, islands, etc., but also
        mereological combinations of these entities) is located at some unique spatial
        region;
CG2.   every spatial region has a unique geographic entity located at it;
CG3.   if two (or more) entities are located are at the same spatial region, then they
        are the same entity.

Consequently, a geography can be considered as non-classical (NCG) if it:

NCG1.   drops one or more of the axioms of CG;
NCG2.   (and/or) adds axioms to those of CG.

According to NCG2, we could, for example, add an axiom to obtain the effect
that all geographic units are connected, or consider how the properties of geographic
boundaries relate themselves to the axioms of classical geography. Instead, regarding
NCG1, we might observe that it licenses non-spatial geographic units as well as maps
with gaps and gluts. To be more precise, denying CG1 also allows the inclusion of
non-spatial geographical entities, entities with multiple location or duplicates of the
same geographical entity. Again, to discard CG2 enables us to consider maps with
regions that are assigned no entity, or two or more competing units.

3.4   Issues from Cartographic Representation

Now, although the distinction between CG and NCG makes no (essential) reference
to maps, Casati et al. (1998) maintain that a model of CG may also be visualized as
a set of instructions for coloring maps, according to which, once a set of colors is
fixed:

3.4.1.   every sub-region of the map has some unique color;
3.4.2.   (and) every color is the color of a unique region of the map.

CG1 would consequently be satisfied by 3.4.2, CG2 by 3.4.1 and CG3 may be
thought as a logical consequence of 3.4.1 and 3.4.2. Analogously, we could easily
generate a CG via an act of tiling which, for example, divides the Earth's surface:

ECG1.   into land and water;
ECG2.   (or) among nations (including quasi-nations such as Antarctica), national
         waters and international waters.

But how to imagine a model of NCG? By circumscribing the analysis to NCG1,
it is interesting to underline that Casati et al. (1998) purpose, in their examples, at
least four different models of NCG.

3.5 The Capital of Singapore

In the first case, a model of NCG is obtained by dropping CG3, according to which if two or more entities share the same location (are located at the same spatial region), then they are the same entity. According to the authors, such a drop would allow maps with spatial regions that are assigned two or more competing locations for geographical entities. As a consequence, this drop would also permit the representation of disputed lands, on which, for example, two (or more) different nations could concurrently declare their sovereignty. The resulting non-classical map could also easily be rethought of in terms that preserve the axioms of CG, for example, by considering all such spatial regions as occupied by geographical entities of the *Disputed Land* type.

However, a more controversial situation might arise by means of another example: that is, by adding to ECG2—the map that divides the Earth's surface into nations, national waters and international waters—some points that locate, on that map, the capitals of all the nations. In this context, on the one hand, we could think that CG3 is not respected. For example, we could consider the points that locate the capitals, as points where two different geographical entities are located at the same time: the nations themselves and their capitals. Otherwise, on the other hand, we may not have difficulties also in considering CG3 as respected. Indeed, we could take all those points in the map as occupied by the (geographical entities) capitals exclusively. As an alternative, we could show how the different conditions of existence and the criteria of identity of nations and (their) capitals do not prevent them from sharing the same spatial location, without creating overlaps—that, in terms of CG, would lead them to be considered as the same entity.

3.6 Looking for No Man's Land

The second model of NCG is achieved by dropping CG2, according to which every spatial region has a unique geographic entity located at it. Besides having maps with two or more competing geographical entities located at the same spatial region, such a drop licenses maps with spatial regions that are assigned no geographical entity. In this context, Casati et al. (1998) remark that a default assignment that preserves the axioms of CG would consist in considering all such spatial regions as occupied by an object of the *No Man's Land* type.

However, the clarifications offered by the authors may not exhaust the issues of the cartographic representation related to the drop of CG2. Let us take, for example, ECG2 and remove the geographical entity *Suriname* from the spatial region occupied by Suriname itself. According to the authors, in this case, we could preserve the axioms of CG by assigning to the spatial region no longer occupied by the geographic entity *Suriname* an entity of the *No Man's Land* type. Despite this, the same result could be obtained by not placing any entity of the kind *No Man's Land* on that spatial region. Indeed, if we keep the distinction between spatial regions and geographical entities, we could do without *Suriname* and *No Man's Land* (entities), only by thinking of *something* whose conditions of existence and identity are simply defined by the boundaries of the neighboring geographical entities[3]—in this case, by French Guyana to the East, Brazil to the South, Guyana to the West and the Atlantic Ocean to the North. At this point, we might further ask whether the boundaries of those geographic entities define a geographic entity or a spatial region. In the former case, it is reasonable to infer that we might have to do with a CG, whereas in the latter with an NCG.

3.7 Sailing to Thule

The last two models of NCG are the outcome of the drop of CG1, according to which every single geographic entity is located at some unique spatial region. In the first of the two models, Casati et al. (1998) consider the possibility of duplicates of the same geographical entity. Such a possibility would contradict CG1, because the entity in question would be located at (at least) two different spatial regions.

The example given by the authors is the People's Republic of China (located in Mainland China) and the Republic of China (located on the Island of Taiwan): both claim to be the *only* China, but we cannot accept both claims if we assume CG1. Now, despite such claims, are we really faced with a duplicate of the same geographical entity, given that the People's Republic of China and Republic of China have different conditions of existence and identity? In other words, are we dealing with a model of NCG that shows duplicates of the same geographical entity? To preserve the axioms of CG, the authors suggest considering *China* (entity) as the mereological sum of the competing spatial regions, which correspond to Mainland China and to the Island of Taiwan. In line with this, may we actually consider (the whole) China as the result of such a mereological sum? To be more precise, could we really define China as the sum of the spatial regions currently occupied by the People's Republic of China and the Republic of China?

Just to add further hurdles, we might also consider the puzzling case of Thule and the several theories about its possible location that include, among others, the coastline of Norway, Iceland, Greenland, Orkney, Shetland, Faroe Islands, and Saaremaa. Now, if we imagine a map that shows all these locations, then we would be hardly inclined to consider Thule as the mereological sum of all the locations ascribed to it. At the same time, it would be unlikely to consider the various Thule(s) represented on the map (with different conditions of identity) as duplicates of the same geographical entity. Perhaps, we could take the various points that locate Thule on that map as indicating different geographical entities, to which different authors have attributed the same reference. But, in that case, how might we interpret that map? As a model of CG or NCG?

---

[3]Cfr. Chap. 5.

3.8 Poland into Exile

The last possibility of providing a model of NCG that excludes CG1 considers the inclusion of non-spatial geographic entities. The example provided by Casati et al. (1998) is that of Poland during the Era of Partition—namely, during the era in which Poland did not have any territory to call its own. According to the authors, such a model of NCG might be converted into a model of CG by naming a certain, arbitrarily chosen, region as *Ersatz*-Poland—for example, the headquarters of the Government in Exile in London—so to preserve its spatial location.

Based on these assumptions, how could we possibly talk about geographical entities such as Kosovo, the Holy Roman Empire, cities, villages or Benelux in ECG2 that, as we said, divides the Earth's surface into nations, national waters and international waters? Considering the non-unanimous recognition by all UN member states, the case of Kosovo is interesting, since its inclusion in ECG2 could depend on its being accepted as an independent nation or as a region that belongs to Serbia. Consequently, if ECG2 did not include Kosovo, such a map could be a model of:

3.8.1.   CG for Serbia, which does not recognize Kosovo as a nation;
3.8.2.   NCG for Italy, which recognizes Kosovo as a nation.[4]

Conversely, if ECG2 included Kosovo, ECG2 could be considered as a model of CG for Italy, but not for Serbia that could see in ECG2 the drop of CG3 for the presence, in the spatial region occupied by Kosovo, of two different competing entities: Kosovo and Serbia. The situation could be further complicated by considering the Holy Roman Empire. Indeed, if we can hardly discard it as a geographical entity, does its exclusion from ECG2 make that map a model of NCG—given the existence of a geographical entity that is currently non-spatial? Or, eventually, should we think about models of CG and NCG only in reference to geographical entities that currently have or can have a spatial location? And if that would be the case, should we also include nations such as the Sahrawi Arab Democratic Republic, the Pridnestrovian Moldavian Republic, and the Republic of Somaliland (nations that have a limited recognition)[5] so to consider ECG2 as a model of CG? And what about geographical entities such as cities and villages? Does the lack of these geographical entities in ECG2 make the entities non-spatial and thus turn ECG2 into a model of NCG? The same question can be extended to geographical entities such as Benelux or to imaginary geographical entities, but also to entities that are the result of a mereological sum of other geographical entities (for example, the mereological sum of New Zealand, Prussia, and Normandy). In all these cases, could the lack of imaginary or arbitrary geographical entities make ECG2 become a model of NCG?

---

[4] Accordingly, Italy might also consider Kosovo as a non-spatial geographical entity, at least in this specific context.

[5] To be more precise, the Sahrawi Arab Democratic Republic is a non-UN member state recognized only by a few UN member states; the Pridnestrovian Moldavian Republic is a non-UN member state recognized only by non-UN member states; the Republic of Somaliland is a non-UN member state not recognized by any state.

3.9   Conclusion

In this chapter, it has been argued that although the distinction between classical and non-classical geographies is a useful tool for specifying the kind of geography implied in the spatial representation, it also reveals some possible ambiguities related to its application to cartographic representation. Those ambiguities make the distinction not entirely clear cut and open up the possibility for different interpretations, which are delegated to the subjects (whoever they are) interested in their application.

A *leitmotiv* of those ambiguities might be identified in the absence of a (shared) definition of geographical entity. Such a *leitmotiv* arises from the four models of NCG and makes the individuation of the kind of geography implied in each spatial representation a difficult task to pursue. Thus, to circumscribe the possible issues related to the geo-ontological debate, it is necessary for us to share a definition of geographical entity that might lay the foundation for an unambiguous application of the distinction between classical and non-classical geographies to the cartographic representation.

So, what is a geographical entity? What are its conditions of existence and identity? Should we include in our rubric of geographical entities only entities that could be portrayed on a map or also non-spatial and/or abstract entities? Should we consider only those entities that currently have a spatial location, or should we also consider those entities that currently do not have one? How do we deal with the geographical entities the location of which is (or was) vague or indeterminate? What is the relation between a geographical entity and the territory it occupies? Can a geographical entity survive or persist without a territory and definite borders? Can it survive or persist with radical changes in its territory or in its borders? Is it essential for it to be where it actually is or to have its actual borders? In Chap. 5, we try to answer these (and other) related questions, dealing with the ontological conundrums generated by the absence of a (shared) definition of a geographical entity.

References

Breysse O, De Glas M (2007) A new approach to the concepts of boundary and contact: toward an alternative to mereotopology. Fundam Inform 78:217–238

Casati R, Smith B, Varzi AC (1998) Ontological tools for geographic representation. In: Guarino N (ed) Formal ontology in information systems. IOS Press, Amsterdam, pp 77–85

Casati R, Varzi AC (1999) Parts and places. MIT Press, Cambridge

Geus K, Thiering M (2014) Common sense geography and mental modelling: setting the stage. In: Geus K, Thiering M (eds) Features of common sense geography. Implicit knowledge structures in ancient geographical texts. LIT, Wien

Mark DM, Smith B, Tversky B (1999) Ontology and geographic objects: an empirical study of cognitive categorization. In: Freksa C, Mark DM (eds) Proceedings of spatial information theory. Cognitive and computational foundations of geographic information science. International conference COSIT'99 stade, Germany, Springer, Berlin, Heidelberg, pp 283–298

Simons P (1987) Parts: an essay in ontology. Clarendon Press, Oxford

Smith B (1994) Fiat objects. In: Guarino N, Pribbenow S, Vieu L (eds) Parts and wholes: conceptual part-whole relations and formal mereology. Proceedings of the ECAI94 workshop, ECCAI, Amsterdam, pp 15–23

Smith B (1995) On drawing lines on a map. In: Frank A, Kuhn W (eds) Spatial information theory—a theoretical basis for GIS. Proceedings, international conference Cosit'95, semmering, Austria (Lecture notes in computer science), vol 988. Springer, Berlin, pp 475–484

Smith B (1996) Mereotopology: a theory of parts and boundaries. Data Knowl Eng 20:287–303

Smith B, Mark DM (1998) Ontology and geographic kinds. In: Poiker TK, Chrisman N (eds) Proceedings of the eighth international symposium on spatial data handling. International Geographical Union, Burnaby, British Columbia, pp 308–320

Smith B, Mark DM (2001) Geographical categories: an ontological investigation. Int J Geogr Inf Sci 15(7), 591–612. http://idwebhost-202-147.ethz.ch/Courses/geog231/SmithMark_GeographicalCategories_IJGIS2001@2005-10-19T07%3B30%3B52.pdf

Smith B, Varzi AC (2000) Fiat and bona fide boundaries. Philos Phenomenol Res 60(2):401–420

Varzi AC (2007) Spatial reasoning and ontology: parts, wholes and location. In: Aiello M, Pratt-Hartmann I, van Benthem J (eds) Handbook of spatial logics. Springer, Berlin, pp 945–1038

Chapter 4
Boundaries

Abstract Chapter 1 argued that the ontological assumptions and commitments of geography include, among others, the notions of geographical entity and boundary. In this chapter, we focus on geographical boundaries, with the aim of analyzing: (1) how the notion of boundary has been conceived by contemporary geo-ontologists; (2) what kinds of geographical boundaries have been identified and categorized; (3) the influence of cultural diversities and human beliefs upon such categorizations. Primarily, we take into account Smith's and Galton's taxonomies as two prominent examples of comprehensive classifications of geographical boundaries. Secondly, starting from Smith and Mark's considerations, the importance of cultural diversities and human beliefs for geo-ontological classifications is discussed. Thirdly, we consider the possibility of the existence of cultural boundaries and the prospect of categorizations that can influence, in their turn, beliefs, culture and individual as well as collective behavior. The idea is to show three different modalities through which culture and beliefs may affect (and have the power to modify) the definition and the individuation of geographical boundaries.

Keywords Boundaries · Classifications · Cultural diversities · Geography · Human beliefs · Ontology of geography

> There was a wall. It did not look important. It was built of uncut rocks roughly mortared. An adult could look right over it, and even a child could climb it. Where it crossed the roadway, instead of having a gate it degenerated into mere geometry, a line, an idea of boundary. But the idea was real. It was important. For seven generations there had been nothing in the world more important than that wall. Like all walls it was ambiguous, two-faced. What was inside it and what was outside it depended upon which side of it you were on.
>
> Ursula K. Le Guin, *The Dispossessed*, 1974.

4.1 The Ontology of Geographical Boundaries

The concept of boundary is a fundamental philosophical issue triggered and required by the reflection upon geography. Its relevance for the geo-ontological[1] debate has been highlighted and studied by different authors,[2] among whom Galton states:

> Boundaries occupy a curiously ambivalent position in any geographical ontology. On the one hand, it seems uncontentious that the primary spatial elements of geography are regions of various kinds: regions are where we live and where things are located. From this point of view, boundaries are only of interest because they define the limits of regions. But precisely because of this, boundaries can acquire a life of their own. The existence of a boundary can have a palpable effect on the behaviour of objects and people in its vicinity. Disputes over territory automatically become focussed into disputes over boundaries, and the boundary itself can become a symbol for the territory it delineates [...]. Indeed, in ordinary speech there is a slippage between 'within this region/area/territory' and 'within these boundaries/limits/borders', pointing to the ease with which we can pass between thinking in terms of regions and thinking in terms of boundaries (Galton 2003, p. 151).

Following Galton's reflection, boundaries embody many different functions. For example, they can regulate motion and/or communication outward from the inside of a region to the outside (inclusion), or inward from the outside to the inside (exclusion). In this sense, a boundary can be built in order to keep someone in or others out, or simply to prevent mixing. But boundaries might be crossed: "physical boundaries such as walls usually include gateways or portals by which movement across the boundary is simultaneously facilitated and regulated. Thus there is another slippage in our thinking, between borders and border-crossings" (Galton 2003, p. 151). Moreover, certain functions such as inclusion and exclusion might be combined (separation), or again there can be an extent to which separation is not complete (contact). Some other functions such as protection are derived from those. Galton also considers that any geographic line can be thought of as a boundary.

> Whether or not it functions as a boundary depends on a variety of factors. As a first high-level generalization, a line can be conceived in two ways: from the point of view of possible motion along it, and from the point of view of possible motion across it. Conceived in the first way, a line is a way or path; in the second, a boundary, barrier or gateway. As Couclelis and Gottsegen (1997) put it, 'a freeway is a way or a barrier depending on which way you look'. Many boundary functions are therefore defined in terms of 'across' rather than 'along'. They have to do with how a boundary regulates movement or communication across it (Galton 2003, p. 163).

But what kind of entities are geographical boundaries? What sorts of boundaries have been identified by contemporary geo-ontologists? How can boundaries be classified from a geo-ontological point of view? What are the main contemporary classifications of geographical boundaries?

---

[1] In this chapter, by "geo-ontological" we specifically mean the area of research at the boundary between analytical ontology and geography; see Chap. 1, Sect. 3.

[2] See for example Febvre (1922), Jones (1963), Prescott (1965), Mark and Csillag (1989), Smith (1995), Burrough and Frank (1996), Smith and Varzi (1997, 2000), Casati et al. (1998), Smith and Mark (1998), Casati and Varzi (1999), Varzi (2016), Piras (2021).

These questions represent the starting point of this chapter, aimed at analyzing:

4.1.1. how the notion of boundary has been conceived by contemporary ontologists of geography;
4.1.2. what kinds of geographical boundary have been identified and categorized;
4.1.3. the influence of cultural diversities and human beliefs upon such geo-ontological classifications.

Given the above, we primarily take into account Smith's (1995)[3] and Galton's (2003) taxonomies as two prominent examples of comprehensive classifications of geographical boundaries, encompassing physical, biological, psychological, social, and political phenomena. Secondly, starting from Smith and Mark's (1998) considerations, the importance of cultural diversities and human beliefs for geo-ontological classifications is discussed. Thirdly, we consider the possibility of the existence of cultural boundaries and the prospect of categorizations that can influence, in their turn, beliefs, culture, and individual as well as collective behavior, showing three different modalities through which culture and beliefs may affect (and have the power to modify) the definition and the individuation of geographical boundaries. The idea behind this chapter is that a study on the influence of cultural diversities and human beliefs on geographical boundaries should not be considered as an exclusive trait of border studies.[4] Rather, it might be something that properly characterizes our way to classify boundaries, also from an ontological perspective.

4.2 Bona Fide and Fiat Boundaries

From a geo-ontological point of view, the first attempt to classify geographical boundaries systematically comes from Smith (1995), who presents a taxonomy of spatial boundaries applied in the areas of geography and of administrative and property law. Such a taxonomy is based on the exhaustive and exclusive distinction between *bona fide* (or physical) and *fiat* (or human-demarcation-induced) boundaries.[5]

Bona fide boundaries (shorelines, riverbanks, coastlines, and so forth) are boundaries in the things themselves, are a matter of qualitative differentiations in the

---

[3] Such a taxonomy has been developed in Smith and Mark (1998), Smith and Varzi (2000), Smith (2019).

[4] Border studies generally understand boundaries as social constructs rather than as naturally given entities. In this respect, see for example Kolossov (2005), Newman (2006, 2010), Agnew (2008), Kolossov and Scott (2013), Paasi (2013a, b), Yachin (2015).

[5] Smith shows three different monistic alternatives to the fiat-bona fide dualism. The first one maintains that all objects are the result of human conceptual articulations and, accordingly, that the idea that there exists an underlying world of bona fide objects is merely the expression of an illegitimate *objectivist* metaphysics. Such a metaphysics presupposes some notion of a "God's eye view" that is held to be inappropriate to our post-enlightenment age. The second alternative considers that no objects are fiat objects and that our talk of the latter is mere talk of no further ontological significance. Finally, the last one maintains that fiat objects are not created but merely selected from the infinite totality of geometrically possible regions of space (Smith 1995, p. 477).

underlying reality, and correspond to genuine discontinuities in the world (Smith 1995, p. 476). Accordingly, they exist even in the absence of all delineating or conceptualizing activities on our part, independently of all human cognitive acts and demarcations.

In contrast, political and administrative boundaries, state and provincial borders, property lines and borders of postal districts provide examples of fiat boundaries, which are delineations that do not correspond to any genuine heterogeneity on the side of the bounded entities themselves. Rather, fiat boundaries exist only in virtue of the different sorts of demarcations effected cognitively and behaviorally by human beings. So, they do not exist independently of human cognitive acts and owe their existence to acts of human decision or fiat, to laws or political decrees or to related human cognitive phenomena. Such boundaries may:

4.2.1.   lie entirely skew to all boundaries of the bona fide sort (e.g., the boundaries of Utah and Wyoming);
4.2.2.   involve a combination of fiat and bona fide portions (e.g., the boundaries of Egypt and Uzbekistan);
4.2.3.   be constructed entirely out of bona fide portions which, however, must be glued together out of heterogeneous portions in fiat fashion in order to yield a boundary that is topologically complete, especially because they are not themselves intrinsically connected (Smith 1995, p. 477).

The exhaustiveness and exclusiveness of the distinction between bona fide and fiat boundaries is neither meant to deny that there are types of spatial boundaries which are difficult to classify under one or another of the two rubrics nor that it may be necessary to introduce a more detailed categorization than this simple dichotomy[6] (Smith 1995, p. 477). According to the last point, Smith and Varzi, for example, introduce some specific fiat boundaries that have a mathematical definition, such as the Equator or the tropics of Cancer and Capricorn. In such cases, "the question of their ontological status is part-and-parcel of the larger question of the existence and status of mathematical entities in reality" (Smith and Varzi 2000, p. 402). Moreover, among these mathematical boundaries, Smith and Mark (1998) also include GIS fiats that are artifacts of a certain technology and that can be considered as specifically scientific fiats.

---

[6]In this sense, according to Smith, cross-cutting this distinction are further oppositions in the realm of boundaries (in particular, of fiat boundaries), for example, between inner and outer, crisp and indeterminate (imprecise, fuzzy or vague), shifting and fixed, complete and incomplete, bi-dimensional and tri-dimensional, enduring and transient, symmetrical and asymmetrical, probable or actual.

4.3   Legal Fiat Boundaries and Normativity

As we said, among fiat boundaries Smith includes boundaries of nations or postal districts, which are social entities, analogous to rights, claims and obligations. To be more precise,

> [There] are fiat boundaries in the social world—such as those drawn by real estate developers or by international boundary commissions—which can be compared to rights, claims, obligations, and other sorts of social object. They have a quasi-abstract character in the sense that they are relatively isolated from causal change. But they are not completely isolated: there is standardly a point in time at which they begin to exist, and while they exist they may be associated with specific systems of legal or other sorts of sanctions. Further, they manifest a type of generic dependence upon associated beliefs and customs on the part of relevant human beings, so that they may be sustained in being from generation to generation (Smith and Varzi 2000, p. 402).

Usually, when the legal system takes a fuzzily bounded region, it has to add a rule (or a norm) to crisp up that boundary. In this sense, another key point for a possible categorization of geographical boundaries can be the analysis of the varieties of normativity behind them. In line with this, it might be useful to examine the consideration proposed by Zaibert and Smith (1999, 2007), who sketch some elements of the ontology of legal and socio-political institutions, by paying attention to the normativity connected to those features. Without claiming to be exhaustive, the two authors maintain that the development of such an ontology must find ways to account for the normative force generated by the normativity of constitutive rules. This means that it is possible to identify at least three varieties of normativity relevant for the ontology of social reality:

4.3.1.   the logically derived normativity "that is closely associated with, if not identical to, the normativity involved in games like chess or poker" (this type of normativity pervades the world of social and legal and political institutions);

4.3.2.   a normativity that is "in no obvious way connected with logic, and which has been the focus of traditional natural law theories" (i.e., "murder is wrong" or "lawmakers ought not to pass laws which conflict with moral obligations");

4.3.3.   the normativity that is "related to the immanent logical structures of mental phenomena and not to conventional games and institutions" (i.e., "to do harm intentionally is more blameworthy than to do harm unintentionally" or "to believe that someone did something blameworthy, is to believe that he ought not to have done it").[7]

However, if we accept the existence of these varieties of normativity, how can we apply them to the geographical boundaries? In other words, how could the identification of, at least, three kinds of normativity behind an ontology of legal and socio-political entities influence the classification of geographical boundaries? Could we really classify geographical boundaries from this normative standpoint? May we

---

[7] See Zaibert and Smith (1999), pp. 17–18.

talk about normative boundaries? If yes, how many and what kinds of normative geographical boundaries may we identify? Or again, may we talk about a normative function of boundaries rather than a specific sub-class of geographical boundaries?

4.4 Physical and Institutional Boundaries

If Smith's classification is primarily focused on the notion of fiat boundaries, the taxonomy of Galton (2003) turns out to be more heterogeneous with bona fide boundaries (or physical boundaries, according to his terminology), presenting some specific sub-classes of them.

Figure 4.1 represents the topmost levels of Galton's categorial hierarchy, which takes the form of a tree structure with a top-level distinction between physical and institutional boundaries. The highest category in the hierarchy is the category of geographical boundaries that, according to Galton, "exist by virtue of the distribution of matter and energy in space and time, but [...] may differ as to just how their existence depends on such distribution" (Galton 2003, pp. 152–3). At the second level, the distinction between physical and institutional boundaries is captured by appealing to the variety of dependence of the boundary on material facts. In the case of institutional boundaries, such a dependence is mediated by individual or collective human intentionality. To be more precise, institutional boundaries are stipulated to exist by human attitudes. In this sense, they include all international and intranational

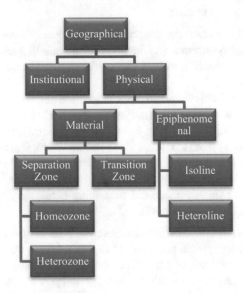

Fig. 4.1 Galton's classification of geographical boundaries

boundaries such as those between administrative regions and those defining land ownership.[8]

All other boundaries are physical ones and are primarily divided into material and epiphenomenal boundaries. In the latter case, boundaries depend on matter for their existence but have no material or phenomenal substance in themselves. In this sense, they exist by virtue of the distribution of matter in space and time but are not themselves made of matter. Conversely, in the case of material boundaries, there is some material substance (or phenomenon) which constitutes the boundary, and the location of the boundary is the location of its material or phenomenal constituents.

Material boundaries are further subdivided in separation zones and transition zones. In both kinds, the boundary occupies a zone whose material or phenomenal contents differ in character from those of the regions on either side. In a transition zone, the character is intermediate between that of one side and that of the other. In a separation zone, the character of the zone is distinct from, and not intermediate between, the characters of the regions it separates (Galton 2003, pp. 153–4). Moreover, separation zones can also be divided in homeozones which separate like from like, and heterozones that separate unlike regions (Galton 2003, p. 154).

Finally, the last subdivision concerns the nature of epiphenomenal boundaries. One kind of epiphenomenal boundary is an isoline for a field, defined as the locus of all points in the field with the same attribute value. The other kind of epiphenomenal boundary is a heteroline, that is any line (or area) of separation between areas of different attribute values.

In conclusion, we have also to underline that, in Galton's opinion, all the distinctions proposed are not (in every case) entirely clear-cut. Indeed, some cases can be classified in different ways depending on how they are interpreted, and we may find intermediate cases which seem to occupy a middle ground between two positions in the classification (Galton 2003, p. 152). Moreover, there can be several cases in which a boundary of one type can evolve into, or otherwise give rise to, a boundary of another type (Galton 2003, p. 159).

4.5 Boundaries, Cultural Diversities, Human Beliefs

As we said, the claim of exhaustiveness of Smith's and Galton's taxonomies should not appear as a restriction for the existence of other kinds of geographical boundaries. On one side, we should consider a certain degree of arbitrariness regarding both what is categorized and how it can be categorized. In this sense, the functions of the boundaries we have to categorize might also assume a significant role. On the other side, we could change the classification system (or propose a new one) and then our boundaries might move, some of them disappear, and new ones might have to be created. Moreover, it is important to remember that the natural language (and its

---

[8] According to Galton, the nature of such boundaries can be clarified by reference to Searle's (1995) theory of institutional facts.

evolution over time) and, more generally, cultural diversities and human beliefs have contributed (and still contribute) to the categorization and the generation of (new kinds of) boundaries. Regarding the last point, the following two quotes from Smith and Mark (1998) may help us to clarify some specific issues.

> Geographic categorization involves a degree of human-contributed arbitrariness on a number of different levels, and it is in general marked by differences in the ways different languages and cultures structure or slice their worlds. It is precisely because many geographical kinds result from a more-or-less arbitrary drawing of boundaries in a continuum that the category boundaries will likely differ from culture to culture (in ways that can lead to sometimes bloody conflict as between one group or culture and another) (Smith and Mark 1998, p. 314).

> Research on this topic must be careful to distinguish the domain of the real world from the domain of computational and mathematical representations, and both of these from the cognitive domain of reasoning, language, and human action. Human practice is an important part of the total ontology. Cultural differences in categorizations are more likely to be found for geographic entities than for objects at table-top scales. Geographic ontologies are more strongly focused on boundaries, and a typology of boundaries is critical. Work involving formal comparisons of geospatial and cartographic data standards and dictionary definitions in a variety of languages will provide an important starting point for the cross-cultural experiments with human subjects that will be needed to refine the details of the ultimate ontology of geographic kinds (Smith and Mark 1998, pp. 317–8).

Without dwelling on the possible influence of Horton's (1982) distinction between primary and secondary theories (and beliefs)[9] on the proposal of Smith and Mark, we could say that such cultural differences might act differently depending on the entities (in our case, boundaries) we want to categorize. On the matter, with reference to bona fide and fiat geographical entities, Smith and Mark assert that:

> bays, peninsulas, etc., are parts of spatial reality, physical parts of the world itself. But they are parts of reality that would not be there absent corresponding linguistic and cultural practices of demarcation and categorization. In a world with our everyday human practices, a bay or a hill is just as real as a chair or rock. The former are real consensus fiat objects, the latter are real bona fide objects. Bona fide objects are for obvious reasons more likely to be objects of categorizations that enjoy a high degree of cross-cultural invariance. Fiat objects, in contrast, because they are inculcated into the world by cognition, are more likely to show cultural dependence (Smith and Mark 1998, p. 315).

Therefore, if we accept such a conclusion or, at least, the idea/conclusion that some geographical boundaries (in particular, according to Smith and Mark, some fiat boundaries) included in our categorizations—or our categorization itself—might be, in some way, culturally influenced, then may we talk about cultural boundaries? And what about the possibility of a categorization of geographical boundaries which (in turn) may have an influence on cultural diversities, human beliefs and individual as well as collective behaviors?

---

[9]See Chap. 2.

4.6 Cultural Boundaries?

To answer the first question, it may be useful to extend our analysis to (some) international and intranational disputes on boundaries that, in my opinion, provide some significant examples of these specific kinds of boundaries. For example, what is the kind of boundary which divides Kosovo and Serbia? Where can we locate the border between Turkey and Kurdistan? Is there a boundary between Padania and the rest of Italy? If we consider the first of the three cases—the border between Kosovo and Serbia—it is easy to imagine how, for Kosovan separatists, this boundary may indicate a dividing-line between two distinct states. Conversely, Serbian nationalists can have a different opinion, considering it as a boundary between two distinct geographical areas, which belong to the same Serbia. The same question can be extended to international affairs, in which Kosovo has been recognized as an independent state by the USA, Canada, and Norway (among others), while, for example, countries such as Argentina, Russia, and Brazil have not. Consequently, the boundary between Kosovo and Serbia can be respectively considered as a national or a regional boundary, depending on the country that draws this specific classification. Moreover, the issue might also be expanded to the spatial representation of the boundary at stake. Indeed, given a map that assigns a different color to each different state, we could easily imagine assigning the same color or two different colors to Kosovo and Serbia, depending on which type of entity Kosovo is taken to be. And if we admit that in these (or other) situations, culture and individual or collective beliefs exert some influences on the classification of boundaries, we could perhaps go further and assert that a classification might also include some boundaries that depend on such factors and give them the name of cultural boundaries.

To answer the second question—that is whether a classification (of geographical boundaries) may (in turn) have an influence on cultural diversities, human beliefs and individual as well as collective behaviors—it might be helpful to recall the declaration provided by the Serbian tennis player Novak Djokovic. After winning the Australian Open 2008, he sent a video message to Belgrade, where 150,000 of his compatriots were demonstrating against Kosovo's declaration of independence. To show his solidarity with the demonstrators, he said: "We are prepared to defend what is rightfully ours. Kosovo is Serbia". [10] I would suggest that in this case Djokovic (as well as the demonstrators) did not intend to propose a classification of geographical boundaries. Rather, he intended to reiterate a pre-existing classification, or better, his position on a classification of a specific geographic boundary. In other words, his declaration (and the transmission of the video message) seems to be the result (and the consequence) of some of his beliefs about that specific boundary. Such beliefs, in turn, could probably be interpreted as the result of various factors and especially of his belonging to a political party that had previously formulated such a classification. Now, accepting such an interpretation probably would lead us to support that the classification proposed by the political party, or better the position expressed by

---

[10]See http://www.spiegel.de/international/europe/street-fighter-artist-and-patriot-tennis-star-djo kovic-is-the-pride-of-new-serbia-a-790484-2.html.

such a political party regarding the categorization of a specific geographic boundary, could have influenced Djokovic in uttering such a declaration and sending the video message. More generally, we could go even further and argue that some classifications of geographic boundaries (or parts of these classifications) might exercise (or even, might be functional to exercise) an influence on beliefs, cultures and behaviors. In line with this assumption, it could be noteworthy to ask whether the specific influence of previous classifications on our system of beliefs is to be taken into account in establishing a classification of geographical boundaries. In other words, does the idea that some boundaries, some types of boundaries or some classifications having an influence—or being specifically created to have one—on beliefs, culture and (also) on individual or collective behavior put forward something new, from an ontological point of view? Should we embrace this aspect in our classification of geographic boundaries? Are we faced with a new type of boundary in which culture is somehow implicated? Or again, are we faced with a new kind (or with a specific sub-class) of cultural boundary? Could we talk of boundaries generated by beliefs on an earlier classification? If this would be the case, is there something that associates and/or distinguishes them from what we have previously identified as cultural boundaries?

4.7   Three Levels of Cultural Dependence

Showing the modalities through which culture, practices and beliefs influence the classification of geographical boundaries (and vice versa) might improperly lead to equate, at least, three different levels on which the influence operates.

Indeed, it is one thing to maintain that the notion of boundary can be culturally determined. Another thing is to say that some kinds of geographical boundaries may show a certain degree of cultural dependence—in particular, according to Smith and Mark (1998), some fiat boundaries and some specific sub-classes of them. Finally, one more thing is to assert that, once a taxonomy (of geographical boundaries) has been accepted, our culture and beliefs can influence us in categorizing a specific boundary (for example, the boundary between Kosovo and Serbia) within one class of boundaries rather than within another (e.g., whether the same boundary might be classified as a national or a regional boundary). In other words, the cultural dependency can occur:

4.7.1.   at the level of the recognition/identification of the notion of boundary;
4.7.2.   at the level of the identification of different kinds of boundaries;
4.7.3.   at the level of the classification of a specific boundary within the different kinds of boundaries previously identified.

Now, although the first of these three levels of cultural dependence might appear, at a first glance, tautological and not (very) informative—we could for example ask what else to add other than the notion of boundary as culturally determined—the considerations of Galton may help to clarify some specific issues. In particular, is the author right in saying that any geographic line can, in principle, be thought of as a boundary? And what if we thought differently? Which criteria could be used to distinguish, among geographical lines, the sub-class of boundaries? In other words, it could be said that what is at issue at this level is the definition of geographical boundary that, *de facto*, determines what can be included in our classification. In this sense, according to various features such as our beliefs, culture, and to our way of considering a definition more compelling than others, we could be willing to accept a certain explanation, to opt for another one, or even to propose a new one. Consequently, the list of entities belonging to our classification of geographical boundaries might change in accordance with the definition we endorse.

Obviously, a reflection on the other two levels might also help to clarify the heterogeneity of the influence of culture, practices and beliefs with regards to the classification of geographical boundaries. But how to distinguish these two levels? Or better, how to discriminate between a kind of boundary and a specific boundary? How can culture, practices and beliefs influence such levels of classification?

An example of a specific boundary, the categorization of which depends on culture and individual and collective beliefs, might be represented by the aforementioned boundary between Kosovo and Serbia. Indeed, without considering further possible alternatives, such a boundary can be regarded as a national or a regional one, according to our beliefs and culture. We can easily see that, in this case, the modality of classification is not called into question, nor is the choice of the kinds of boundaries identified. Instead, what is involved is only the choice whether to include this specific boundary in a particular sub-class, rather than in another one.

Finally, as a possible example of a kind of boundary the identification of which can be influenced by culture and individual as well as collective beliefs, we could, for example, examine a specific class of fiat boundaries provided by Smith: property boundaries. Indeed, if we have a little doubt in accepting the fact that a wall, a hedge, but also an imaginary line can constitute (of course, not necessarily) a boundary between two different properties, we could hardly imagine the same in a society that does not know or has never known (or also that does not accept) the concept of property. Within such a society, although entities such as walls, hedges, and so forth can still be considered as boundaries (according to the aforementioned reflection of Galton: "any geographic line can be thought of, in principle, as a boundary"), they might not, *de facto*, be considered as property boundaries, because this concept does not exist in that society. The same line of reasoning might easily be extended to national borders before the birth of the concept of nation, to GIS boundaries before the development of these technologies, and so forth.

4.8 Conclusion

The aim of this chapter has been twofold. On the one hand, it has shown how the notion of geographical boundaries has been classified from a geo-ontological point of view. On the other hand, some perspectives regarding the possible influence of culture, practices and (individual and collective) beliefs on the modalities of classification

of these boundaries have been provided. Firstly, it has discussed the possible existence, among the geographical boundaries included in our classifications, of cultural
boundaries, i.e., boundaries of which the recognition and location are, in some way,
influenced by beliefs and/or culture. Then, we have discussed the possibilities of
categorizations of geographic boundaries that can influence beliefs, culture, and
individual or collective behavior in turn. Finally, three different modalities through
which culture and beliefs with an influence and the power to modify the definition and
the individuation of geographical boundaries have been distinguished and presented.

Obviously, the proposed analysis is not intended to be exhaustive. First of all, it
does not exclude the existence of other possible cases in which culture and beliefs
might have a grip on our classification of geographic boundaries. Second, what has
been proposed is limited to the analysis of fiat cultural boundaries and has excluded
the possible influence of culture and beliefs on the classifications of bona fide/physical
boundaries—but also on the distinction between physical or bona fide boundaries
and institutional or fiat boundaries itself. Third, with a remark on the preliminary
nature of this reflection, it should be stressed that the possible inclusion of cultural
boundaries in our classifications might lead to a series of issues related to precisely
tracing the distinction between what can be classified as cultural, non-cultural or not
entirely cultural. In other words, paraphrasing the words of Galton, the distinction
here proposed is not entirely clear-cut and some cases can be classified in different
ways depending on how they are interpreted. On the one hand, we may find intermediate cases which seem to occupy a middle ground between the various kinds of
cultural influence proposed. On the other hand, there can also be several cases in
which a cultural boundary of one type can evolve into or otherwise give rise to a
boundary of another type and vice versa.

References

Agnew J (2008) Borders on the mind: re-framing border thinking. Ethics Global Pol 1(4):175–191
Burrough PA, Frank AU (eds) (1996) Geographic objects with indeterminate boundaries. Taylor &
    Francis, London
Casati R, Smith B, Varzi AC (1998) Ontological tools for geographic representation. In: Guarino
    N (ed) Formal ontology in information systems. IOS Press, Amsterdam, pp 77–85
Casati R, Varzi AC (1999) Parts and places. MIT Press, Cambridge
Couclelis H, Gottsegen J (1997) What maps mean to people: denotation, connotation, and geographic
    visualization in land-use debates. In: Hirtle SC, Frank AU (eds) Spatial information theory: a
    theoretical basis for GIS, vol 1329 of lecture notes in computer science. Springer, Berlin, pp
    151–162
Febvre L(1922) A geographical introduction to history (eng. Trans by Mountford E, Paxton JH
    (1925). Knopf, New York
Galton A (2003) On the ontological status of geographical boundaries. In: Duckham M, Goodchild
    MG, Worboys MF (eds) Foundation of geographic information science. Taylor & Francis, London,
    New York, pp 151–171
Horton R (1982) Tradition and modernity revisited. In: Hollis M, Luke D (eds) Rationality and
    relativism. Blackwell, Oxford, pp 201–260
Jones S (1963) Weights and measures: an informal guide. Public Affairs Press, Washington, D.C.

Kolossov V (2005) Border studies: changing perspectives and theoretical approaches. Geopolitics 10(4):606–632

Kolossov V, Scott J (2013) Selected conceptual issues in border studies. Belgeo 1. http://belgeo.rev ues.org/10532

Le Guin UG (1974) The dispossessed: an ambiguous Utopia. Harper & Row, New York

Mark DM, Csillag F (1989) The nature of boundaries on 'area-class' maps. Cartographica 26:65–78

Newman D (2006) The lines that continue to separate us: borders in our 'borderless' world. Prog Hum Geogr 30(2):143–161

Newman D (2010) Territory, compartments and borders: avoiding the trap of the territorial trap. Geopolitics 15:773–778

Paasi A (2013a) Borders and border crossings. In: Johnson N, Schein R, Winders J (eds) A new companion to cultural geography. Wiley-Blackwell, London, pp 478–493

Paasi A (2013b) Borders. In: Dodds K, Kuus M, Sharp J (eds) The Ashgate research companion to critical geopolitics. Ashgate, London, pp 213–229

Piras N (2021) The metaphysics of boundaries. Springer, Cham

Prescott JRV (1965) The geography of frontiers and boundaries. Hutchinson, London

Searle JR (1995) The construction of social reality. Penguin, Harmondsworth

Smith B (1995) On drawing lines on a map. In: Frank A, Kuhn W (eds) Spatial information theory— a theoretical basis for GIS. Proceedings, international conference Cosit'95, Semmering, Austria. Lecture Notes in Computer Science, vol 988. Springer, Berlin, pp 475–484

Smith B (2019) Drawing boundaries. In: Tambassi T (ed) The philosophy of GIS. Springer, Cham, pp 137–158

Smith B, Mark DM (1998) Ontology and geographic kinds. In: Poiker TK, Chrisman N (eds) Proceedings of the eighth international symposium on spatial data handling (Burnaby, British Columbia, International Geographical Union), pp 308–320

Smith B, Varzi AC (1997) Fiat and bona fide boundaries: towards an ontology of spatially extended objects. In: Hirtle SC, Frank AU (eds) Spatial information theory: a theoretical basis for GIS, vol 1329, lecture notes in computer science. Springer, Berlin, pp 103–119

Smith B, Varzi AC (2000) Fiat and bona fide boundaries. Philos Phenomenol Res 60(2):401–420

Varzi AC (2016) On drawing lines across the board. In: Zaibert L (ed) The theory and practice of ontology. Palgrave Macmillian, London, pp 45–78

Yachin SE (2015) Boundary as an ontological and anthropological category. In: Sevastianov SV, Laine JP, Kireev AA (eds) Introduction to border studies. Dalnauka, Vladivostok, pp 62–79

Zaibert L, Smith B (1999) Legal ontology and the problem of normativity, the analytic-continental divide. Conference, University of Tel Aviv

Zaibert L, Smith B (2007) The varieties of normativity: an essay on social ontology. In: Tsohatzidis SL (ed) Intentional acts and institutional facts. Essays on John Searle's social ontology. Springer, Dordrecht, pp 157–173

Chapter 5
Geographical Entities

Abstract As one of the main aims of applied ontology of geography is to establish what kinds of geographical entities exist, this chapter provides a sketch of possible approaches, response attempts, and issues arising from the question: "What is a geographical entity?". It is shown how trying to answer this question is made particularly difficult by a multiplicity of aspects that might be summarized as follows: (1) There exist multiple conceptualizations of the geographical world. (2) Different languages and cultures slice such a world in different ways. (3) The geographical world has changed and will change over time. (4) Geography (as a discipline) has changed and will change over time too, by modifying its perspectives, tools, and domains of investigation. Consequently, what was, is, and will be considered as non-geographic can be considered as geographic and vice versa. (5) There were, are, and will be different kinds of geographies as well as different geographical branches, each of which had, has, and will have different tools, aims, and vocabularies. (6) The introduction of new scholarly fields and new technologies, the birth of intellectual movements or paradigm shifts can/will influence geography as a discipline.

Keywords Boundaries · Definitions · Geographical entity · Hierarchical structures · Kinds · Laundry lists · Maps · Perspectivism · Relations · Vagueness

5.1 A Chaotic List that Cries Out for Explanation

Providing a full list of geographical entities could be a time-consuming and (potentially) extravagant objective due to the innumerable functions that geographical entities might have and the variety of (disciplinary) contexts from which they emerge.

They might arise from the physical world such as, for example, mountains, seas, oceans, rivers, islands, deserts, and so forth (Inkpen and Wilson 2013). They can emerge from the combination of environmental features (of the physical world) and demarcations introduced by human cognition and action (i.e., bays, promontories, and so on). They might also be the result of political and administrative subdivisions, law decrees, land ownerships such as nations, regions, postal districts, and so on, involving social conventions on several different levels, generally marked by differences in the ways different societies structure the world (Smith and Mark 1998). In addition, an inventory of geographical entities can also include human-made objects such as streets, buildings, and so forth (Laurini 2017).

Obviously, we could go on and on, listing kinds and sub-kinds of geographical entities or emphasizing that they may be zero-, one-, two-, or three-dimensional such as, respectively, the South Pole, the Tropic of Cancer, Canada (a two-dimensional object with a curvature in three-dimensional space) and the North Sea—that, according to the context, can refer either to the three-dimensional body of water, or to its two-dimensional surface (Smith and Mark 1998).

Moreover, geographical entities can be disconnected like countries with several islands or like France with Martinique, Guyana, New Caledonia, etc. Sometimes, they have *holes* such as South Africa has with Lesotho. Some have sharp borders, others indeterminate boundaries. They also can be simple, made up by other geographical entities, or share mereological or topological relations with other geographical entities or/with their parts (Varzi 2007).

Furthermore, some geographical entities also have some sorts of relations both with the (surface of the) Earth and with the space they occupy (Casati and Varzi 1999). Generally speaking, a relational theorist of space would say that entities are cognitively and metaphysically prior to space (there is no way to identify a region of space except by reference to what is or could be located or take place at that region). In contrast, an absolutist theorist would say that space exists as an independently subsistent individual (a sort of container) over and above its inhabitants (objects, events, and spatial relations between objects and events or without all these entities).

This chaotic list cries out for order and explanation. Is there really something such as a geographical entity? What, if anything, do geographical entities have in common? What sorts of entities are they, how are they individuated, and what are their identity conditions (Bishr 2007)? How do we distinguish between what is a geographical entity and what is not? What is the difference, if any, between geographical and spatial entities? Are there geographical entities that are not spatial and/or spatial entities that are not geographical? What are the sorts of factors that might influence our inventory of geographical entities? What is the relationship between geographical entities and their representation on maps (Casti 2015)? Should we think of geographical entities in general, or is it more appropriate to assume that every geographical sub-area has a proper list/account of geographical entities?

5.2   Avoiding Univocal and Incomplete Accounts

In approaching those questions, one of the main issues will be to avoid univocal and incomplete accounts, which could be suitable for some theoretical tasks but not rich enough to grasp the complexity of our ways of representing the geographical world. Geography, indeed, has had hundreds of years to elaborate different sorts of geographical entities for innumerable purposes. Moreover, geography (as a discipline) has changed over time, modifying its perspective, sometimes its aims, subdividing itself in different branches, and weaving together with different scientific, social, and technical disciplines (Pattinson 1963; Couclelis 1998; Bonnett 2008; Sala 2009). In this sense, what can be considered as geographic has changed according to the geographical perspective we endorse. Furthermore, we should observe that different languages and cultures have created different vocabularies and ways of slicing the world, producing (potentially) different kinds of geographical entities. The fact that geographical reality/realities can be studied, sliced, and represented in different ways does not exclude that such alternative descriptions of the geographical world can be compared, overlapped, and/or integrated with one another in order to get hold of improved accounts of reality itself.[1] Hence, by paraphrasing the words of Epstein (2017), the proposed analysis will be multifaceted and will fight the prevailing philosophical trend of simplifying the endless diversity and variation among different geographical perspectives (Elden 2009; Tanca 2012; Günzel 2015).

In light of these considerations, this chapter presents a series of possible issues, conundrums, and approaches for analyzing and explaining the nature of geographical entities. Surely, such a series should not be conceived as exhaustive, nor the approaches as isolated one from another. Instead, there can be mixed cases that might be seen as a combination of different approaches—for example, between laundry lists, attempts of definition, and others. The same can be said both for issues and conundrums, which rarely appear alone and sometimes also persist across the different approaches we discuss.

Finally, we should underline that the choice of the term "entity" is not neutral in this context and can be considered as problematic. Indeed, as Smith and Mark (2001) remark, philosophical ontologists have long been aware of the controversial character of ontological terminology. In this sense, the term that should be used for the ontological supercategory (things, entities, items, existents, realities, objects, some-things, tokens, instances, particulars, individuals) within which everything belongs is not exempt from possible criticisms. Each alternative has its adherents, yet each also brings problems and sometimes different inventories. In this case, the choice of "entity" is given by the needs of generality and exhaustiveness (i.e., common to other terms), which are specified by the possibility of being inclusive, on the one hand for items such as relations, kinds and so forth, and on the other hand for things that might be abstract, universals, or non-instantiated. Accordingly, such a choice means to not exclude, a priori, the possible existence of these sorts of things, which could be easily compromised, for example, by the use of terms such as existents, particulars, instances and so on. With this, I am not saying that using terms other than "entity" is not appropriate, nor that it cannot bring (in principle) similar results. Rather, I would like to remark that the purpose of this chapter is to show some different approaches, issues, and conundrums which emerge from the debate on geographical entities, with the aim of drawing a boundary (at least, a theoretical boundary) for distinguishing what is geographical from what is not.

---

[1] Therefore, my theoretical point of view may be seen as closed to ontological perspectivism; see Bateman and Farrar (2004), Grenon and Smith (2007), Elford (2012).

5.3   Laundry Lists

Within the philosophical debate, when asked to provide a definition of "ontological category",[2] a possible answer consists in giving not some particular examples of ontological categories, but a full list of all of them, without further specifications. Surely, it is useful to know what has been regarded as an ontological category in the history of philosophy or what a particular author regards as such. But no matter how interesting a list might be in itself, it is certainly no substitute for a definition. Rather, the list sets the stage by indicating which kinds of things our definition should incorporate (Westerhoff 2005, pp. 23–4).

Now, if the narrow number of ontological categories should guarantee an almost exhaustive list of ontological categories,[3] a list of geographical entities can with difficulty strive for such an exhaustiveness. Therefore, a laundry list position in geography will give only a few (and maybe paradigmatic) examples of geographical entities, at the expense of a long and tedious catalog of all of them. But how to provide such examples? According to Varzi (2001), normally, we know how to use geographic terms without being able to provide a precise explanation of the grounds for this competence. Presumably, the model of family resemblance shows how, in ordinary circumstances, a word can be used successfully regardless of whether or not it meets the Fregean ideal of precision. We say that something is a geographic entity because it resembles (in some relevant respect) several things that have hitherto been said to be a geographic entity, even if the exact nature of this resemblance may give rise to borderline cases.

Nowadays, the laundry list position constitutes a recurring integration of many attempts at definition of geographical entity.[4] A non-exhaustive explanation can be traced in the ambiguous epistemological status of geography (that ranges, among others, from physical and human approaches to the spatial analysis), for which a laundry list, even before a definition, seems to guarantee a continuity among different geographical sub-areas. However, some difficulties may arise in deciding whether some particular entities in the lists are (or not) geographical entities. In that case, what criteria should we use for selecting the geographical entities from the realm of entities? What unifies a nation, a mountain, a latitude and makes us classify them as geographical entities? In other words, what, if anything, do geographical entities have in common?

---

[2] See Chap. 7.

[3] In theory, such a list should also be open to the inclusion of new empirical and theoretical evidence, which might modify and/or extend it.

[4] See for instance: Casati et al. (1998), Smith and Mark (2001), and the link https://definedterm.com/geographic_entity. In the geographical debate, other examples of laundry lists (that integrate some attempts to define "geographical entity") can also be found in some of the more general classes of geo-ontologies. For a list (not a laundry list) of the main geo-ontologies, see Chap. 6.

5.4 Attempts of Definition

A possible answer to the questions above might deal with the *definition of geographical entity*—that is, to specify what a geographical entity is by exhibiting its conditions of existence, individuation and persistence and its criteria of (synchronic and diachronic) identity. According to Bishr (2007), identity criteria provide sufficient conditions for determining the identity of entities defined in the domain we have to describe. For the purpose of this chapter, providing identity criteria might be useful for:

5.4.1. classifying an entity as an instance of the class "geographical entity" [GE];
5.4.2. individuating an entity as a countably distinct instance of GE;
5.4.3. identifying two entities at a given time (synchronic identity);
5.4.4. re-identifying an instance of GE across time (persistence or diachronic identity).

Once we fix the identity criteria for geographical entities, it is essential to determine what (geographic) entities (objects, relations, kinds, facts, events and so forth) have to be included as fundamental. Moreover, we should establish whether our list of geographical entities comprehends only entities, such as mountains, rivers, and deserts, traditionally linked to the physical geography and/or also artifacts studied by human geography (entities like socioeconomic units, nations, cities, and so on). In this regard, Casati et al. (1998) have distinguished three main positions on the existence of geographic entities:

5.4.5. strong methodological individualism—there are "only people and the tables and chairs they interact with on the mesoscopic level, and no units on the geographic scale at all";
5.4.6. geographic realism—"geographic entities exist over and above the individuals that they appear to be related to and have the same ontological standing as these";
5.4.7. weak methodological individualism—if geographic units exist, "then they depend upon or are supervenient upon individuals. One form of this position would accept both individuals and the behavioural settings in which individuals act. Larger-scale socio-economic units would then be accounted for in terms of various kinds of connections between behavioural settings" (Casati et al. 1998, p. 79).

However, despite these clarifications, some issues remain unaffected. What entities should we classify as instances of the class "geographical entities"? How do we distinguish between what is a geographical entity and what is not? What is the definition of geographical entity? How could we possibly distinguish, among the physical entities, those that are specifically geographic? Where is the exact boundary between physical and human (geographical) entities? And between spatial and geographical entities?

5.5   On Being Portrayed on Maps

> It seems that [...] it is being portrayable on a map which comes closest to capturing the meaning of "geographic" as this term is employed in scientific contexts. Geographers, it seems, are not studying geographical things as such things are conceptualized by naïve subjects. Rather, they are studying the domain of what can be portrayed on maps (Smith and Mark 2001, p. 609)

Smith and Mark have made the point clear: if geographers study the domain of what can be portrayed on maps, then being portrayed on maps can say something about the notion of geographical entity. Now, let us suppose that the notions of "being portrayed on maps" and "geographical entity" correspond—in other words, that:

5.5.1.   an entity is geographical if and only if it can be portrayed on maps;
5.5.2.   something can be portrayed on maps if and only if it is a geographical entity.

Obviously, such an identity relation could easily solve some problems concerning the identification of geographical entities. However, it might also raise many ontological conundrums that have not yet been addressed.

The first one is that the question of the definition seems to be simply shifted from the notion of geographical entity to the notion of map. This might also lead to the subordination of the notion of entity to its representation[5] and maybe, more generally, the geography to the cartography.[6]

The second conundrum concerns what to do in the face of:

5.5.3.   non-spatial geographical entities, which can be difficult to locate on a map— for example, Poland during the Era of Partition, that is, the era in which Poland did not have any territory to call its own;
5.5.4.   unusual maps, for instance treasure maps, maps that also include imaginary entities (such as Atlantis), maps using GPS coordinates (such as Google Maps), and so forth.

In all these cases, would we be willing to include the treasure, Atlantis, and/or maybe ourselves in the inventory of geographical entities, too?

Finally, the third conundrum is strictly related to the first one and concerns the relationship between the notions of map and geographical entity. If, on the one side, we can have some difficulties in imagining geographical entities that cannot be portrayed on a map, on the other side, some issues might arise with extremely detailed maps that represent not only entities such as seas, nations, and streets, but also trees, sidewalks, and lampposts, which we would probably have more trouble classifying as properly geographic. But then, what would we be willing to include among the geographical entities?

---

[5] In which case, we should, perhaps, also ask whether it is more appropriate to talk about cartographic entities rather than geographical entities.

[6] About the non-correspondence between geography and cartography, and more in general on the critique of the "cartographic reason", see Farinelli (2003, 2009).

5.6 Maps, Granularity of Interest, and Multiple Levels of Details

A possible way of answering the previous question might be to include, among the geographical entities, also entities like trees, sidewalks, and lampposts. But in the face of more detailed maps, the risk is that our list of geographical entities also comprehends the leaves of those trees, the columns of those lampposts, a blade of grass of a garden, and so forth. An alternative can be to consider (only) maps which are not so detailed, that is, maps containing only geographical things. But, how to build such maps?

One of the issues in this matter might be the concept of granularity of interest, according to which geographic objects can mutate in two different ways:

5.6.1. as the scale diminishes, an area mutates into a point and then disappears;
5.6.2. as the scale enlarges, something might appear as a point and then mutate into an area.

Of course, a conceptualization of geographic space may have several levels of granularity, each of which have a specific inventory of geographical entities at different levels of detail. However, nothing excludes that once the scale is enlarged, the granularity of interest might also contain manipulable objects (see Sect. 5.7) that, according to Egenhofer and Mark (1995), should not be properly included in the geographic space (and within geographical entities). Rather, the two authors maintain that geographic space shall include entities such as "hotel with its many rooms, hallways, floors", "Vienna, with its streets, buildings, parks, and people", "Europe with mountains, lakes and rivers, transportation systems, political subdivisions, cultural variations, and so on". In other words, geographical space represents the space in which we move around and that may be conceptualized from multiple views, which are put together (mentally) like a jigsaw puzzle. To put it differently, it is the level of granularity that coincides with the mesoscopic stratum of spatial reality. (The other stratum is the microphysical one that may be conceived as a complex edifice of molecules). The mesoscopic stratum is the real-world counterpart of our non-scientific cognition and action in space and has three different types of components:

5.6.3. objects of a physical sort (such as rivers, forests, seas) that are studied also by physics but which, within the mesoscopic stratum, have different sorts of properties—this is in virtue of the fact that our naïve cognition endows its objects with qualitative rather than quantitative features and with a social and cultural significance that is absent from the microphysical realm;
5.6.4. objects like bays and promontories, which are also in a sense parts of the physical world but exist only in virtue of demarcations induced by human cognition and action;
5.6.5. geopolitical objects such as nations and neighborhoods that are more than physical, and which exist only as the hybrid spatial products of human cognition and action (Smith and Mark 1998, p. 313).

However, despite these clarifications, some issues remain unsolved. Which maps should we properly refer to? What is the minimum level of granularity for a map that represents exclusively geographic entities? What is the difference between geographic and manipulable objects? Such questions seem to reveal some limits of the correspondence between the notions of "being portrayed on maps" and "geographical entity" and highlight a sort of primacy of the latter notion (or, at least, of the evaluation of what is properly geographic) over the notion of "being portrayed on maps". But then, how to distinguish between what is a geographical entity from what is not?

5.7 On What and Where

The theory of spatial location investigates the relation between geographical entities and the regions of space they occupy or in which they are located. According to Casati et al. (1998), specifying such a relation also means choosing, in terms of representation, between classical and non-classical geographies.[7]

Classical geography assumes that every single geographic entity is located at some unique spatial region and that every spatial region has a unique geographic entity located at it. Consequently, classical geography defines the relation between geographical entities and the regions they occupy in terms of identity. As stated by Bishr (2007), such an identity relation also constitutes a fundamental premise of GIS and geo-ontologies, according to which a (geographical) object must have some location, even if the location can be arbitrary. In contrast, non-classical geographies consider that the relation in question is not one of identity. That implies the possibility of geographical entities that are not located somewhere, of spatial regions with two or more geographical entities located at them and/or without entities on them. In other words, non-classical geographies license:

5.7.1.   on the one hand, non-spatial geographical entities, entities with multiple location, or duplicates of the same geographical entity;

5.7.2.   on the other hand, maps with regions that are assigned no entity, or two or more competing units.

By discussing the relation between what and where, the theory of spatial locations also allows us not to consider geographical objects as larger versions of the everyday objects and kinds studied in cognitive science. Indeed, according to Smith and Mark (1998), geographic objects are not merely located in space, as are the manipulable objects of tabletop space or roughly human scale such as birds, pets, toys, and other similar phenomena. For the latter, the "what" and the "where" are almost independent. On the contrary, in the geographic world, the "what" and the "where" are intimately intertwined. To be more precise, geographical objects are tied intrinsically to space, in a manner that implies that they inherit from space many of

---

[7]See Chap. 3.

its structural (mereological, topological, geometrical) properties. Obviously, that is
not the only difference. According to the authors, geographic reality comprehends
mesoscopic entities, many of which are best viewed as shadows cast onto the spatial
plane by human reasoning and language (and by the associated activities). Because
of this, geographic categories are much more likely to show cultural differences in
category definitions than are the manipulable objects of tabletop space. Furthermore,

> In the geographic world, categorization is also very often size- or scale-dependent [...]. In
> the geographic world, to a much greater extent than in the world of table-top space, the real-
> ization that a thing exists at all may have individual or cultural variability. In the geographic
> world, too, the boundaries of the objects with which we have to deal are themselves salient
> phenomena for purposes of categorization. [...] Moreover, the identification of what a thing
> is may influence the location and structure of the boundary. (Smith and Mark 1998, p. 309)

5.8 Drawing the Contour

Another strategy for the identification and the individuation of (autonomous)
geographical entities starts with the specification of their boundaries, in terms of location and typology. To be more precise, the strategy consists in sketching a taxonomy of
boundaries, from which it may derive a corresponding categorization of the different
sorts of geographical entities that boundaries determine and/or demarcate.[8] The basic
idea is that an analysis on (and a classification of) geographical boundaries might
be functional for determining what kinds of geographic entities exist and have to be
included as fundamental.

As stated in Chap. 4, Smith (1995) and Galton (2003) have provided two prominent examples of comprehensive classifications of geographical boundaries for the
geo-ontological domain. Both the classifications take the form of a hierarchical tree
structure with a top-level distinction, which is considered, by the authors, as absolute, exhaustive, and mutually exclusive. Galton distinguishes between institutional
and physical boundaries. Such a distinction is the result of the different distribution
of matter and energy in space and time, from which the existence of boundaries
depends. For institutional boundaries, the dependence of the boundary on the material facts is mediated by individual or collective human intentionality. For physical
boundaries, there is not such a meditation. Conversely, Smith's main distinction is
between *bona fide* and *fiat* boundaries. Bona fide boundaries exist even in the absence
of all delineating or conceptualizing activity on our part. Therefore, they exist independently of all human cognitive acts and demarcations. On the contrary, the existence of the fiat boundaries depends on our delineating or conceptualizing activity.
Despite Galton and Smith not sharing the same terminology, the examples they use
for the entities belonging to such categories seem to indicate an overlap between
the distinctions above. Indeed, both the authors include entities such as coasts, riverbanks, seaboards, and so forth among the prototypical examples of physical/bona fine

---

[8] See Smith and Varzi (1997), Casati et al. (1998), Smith and Mark (1998), Smith and Varzi (2000),
Galton (2003).

boundaries. In contrast, entities like political and administrative boundaries, state and provincial borders, property lines and borders of postal districts provide examples of institutional/fiat boundaries.

Now, the main issue arising from this strategy concerns whether we can really affirm that the notion of boundary is, in some way, prior to the notion of geographical entity. If no, we might, in principle, assume the existence of geographical entities without boundaries. Consequently, such a position can hardly be considered as exhaustive in providing a complete inventory of geographical entities. If yes, we should analyze the ontological status of (geographical) boundaries and (maybe) choose whether:

5.8.1.   to consider them as higher-order entities, as some eliminativist theories do (see Sect. 5.16);

5.8.2.   or, conversely, to include them within the list of geographical entities (as mountains, rivers, cities are).

Both 5.8.1 and 5.8.2 require an explanation of how a class (or a subclass) of geographical entities can play a normative role in the definition of such entities, avoiding a *petitio principii*.

Another issue might also arise from the claim of exhaustiveness of the taxonomies above, which should not appear as a restriction for the existence of other kinds of geographical boundaries. On one side, we should consider a certain degree of arbitrariness regarding both what is categorized and how it can be categorized. In this sense, the functions of boundaries that we want to categorize might also assume a significant role. On the other side, we could also change the classification system (or propose a new one) and then some of our boundaries might move, some of them disappear, and new ones might have to be created. Moreover, it is important to remember that the natural language (and its evolution over time) and, more generally, cultural diversities in addition to human beliefs have contributed (and still contribute) to the categorization and the generation of (new kinds of) boundaries.

Finally, paraphrasing the words of Galton, even the distinctions proposed may not entirely be clear-cut and some cases can be classified in different ways depending on how they are interpreted. On the one hand, we may find intermediate cases, which seem to occupy a middle ground between two positions in the classification. On the other hand, there can be several cases in which a boundary of one type can evolve into or give rise to a boundary of another type and vice versa (Galton 2003, pp. 159, 152).

5.9 Cultural Entities

As Smith and Mark (1998) have remarked, we should also consider that geographic entities (and, more in general, geographic subdivisions) might involve a degree of human-contributed arbitrariness, which is generally marked by differences in the

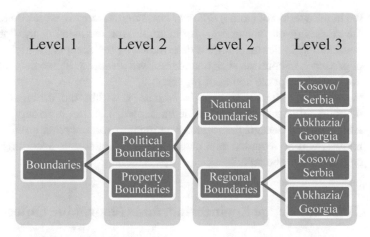

Fig. 5.1 Levels of cultural dependency

ways different languages (and their evolutions over time), beliefs and, in particular, cultures structure or slice the geographical world.[9] According to the authors, such (cultural) differences can act differently depending on the entities we want to categorize:

5.9.1.   bona fide entities (seas, mountains, lakes, deserts) are more likely to be objects of categorizations that enjoy a high degree of cross-cultural invariance;

5.9.2.   fiat entities (nations, provinces, postal districts), in contrast, as far as they are inculcated into the world by cognition, are more likely to show cultural dependence.

Accepting that some geographical entities (included in our categorizations) might be, in some way, culturally influenced may leave the door open to the introduction of:

5.9.3.   cultural (geographical) entities in our classifications;

5.9.4.   categorizations of geographical entities which (in turn) may have an influence on cultural diversities, human beliefs, and individual or collective behaviors.

Regarding 5.9.3, we should also consider that the modalities through which cultural differences might influence the classification of the geographical entities (and vice versa) operate, at least, at three different levels that should not be (improperly) equated. To be more specific (see Fig. 5.1), by using the notion of geographical boundaries, we can specify that cultural dependency can occur, at least, at the level of:

5.9.5.   the *definition of geographical boundaries* that determines what should be included in (the full list of entities belonging to) our classification;

---

[9] See Chap. 4.

5.9.6.   the identification of (some) *different kinds of boundaries*, which determines the classes of our taxonomy—for example, the inclusion of the "property boundaries" in our taxonomy is determined by the acceptance of the notion of property. In contrast, such boundaries will disappear in a society that does not know/accept the notion of property;

5.9.7.   the *categorization of a specific boundary* within the different classes of boundaries previously identified/accepted; i.e., the boundary between Abkhazia and Georgia that, without taking into account other possible alternatives, can be regarded as a national or a regional one, according to our cultures and/or beliefs.[10]

5.10   GIS, Knowledge Engineering, and Geographic Objects

If most of the approaches and considerations above generally adopt a speculative viewpoint, the perspective of Laurini (2017) is just to describe the notion of geographical object within the domain of GIS and knowledge engineering. In this context, the author maintains that any geographic object should have:

5.10.1.   an ID named "GeoID", which is an identifier only used for storing;
5.10.2.   a geographic type;
5.10.3.   a geometric shape (the most accurate possible)—and when necessary other less accurate representations are derived by using generalization algorithms;
5.10.4.   zero, one or many different toponyms;
5.10.5.   links with other geographical objects by spatial or geographic relations or even by structures.

Such a list extends and specifies the three different facets that, according to the author, characterize the peculiarity of geographical objects within the domain of GIS and knowledge engineering: geometry, identification, and semantic.

By their geometric facet, Laurini distinguishes two main categories of geographical objects:

5.10.6.   crisp objects that have well-defined boundaries, such as administrative objects (countries, regions, natural parks, etc.) and man-made objects (streets, buildings, and so forth);

---

[10]A similar example is provided by the recognition of Kosovo and the Pridnestrovian Moldavian Republic that is supported only by some of the members of the United Nations. In other words, some members of UN consider Kosovo and the Pridnestrovian Moldavian Republic as proper nations, while other members do not. Consequently, the categorization of these entities changes according to the member of the UN that classifies them. Of course, the concept of recognition is neither a prerogative of the United Nations nor of the notion of nation. On the contrary, it may be applied to, in principle, other geographical notions and/or institutions. See, for example, Italy with Lunezia in Sect. 5.17.

5.10.7.   fuzzy objects, which have undetermined boundaries (mountains, marshes, deserts, etc.).

The first category of geographical objects (5.10.6) might be represented by using conventional geometry (that should also take into account issues coming from the curvature of the Earth), whereas the second one (5.10.7) requires models deriving from fuzzy sets.

From the point of view of identification, Laurini maintains that geographic objects can have names, sometimes several names, and that the same name might also be assigned to several entities. The introduction of gazetteers and computer identifies (IDs) allows us to solve some ambiguities arising from toponymy, even though in different databases the same features can have different identifiers.

Finally, due to their semantics, geographical objects might be considered as conventional objects. However, some issues can emerge from the fact that different languages might:

5.10.8.   confer different names to the same geographical entity (e.g., Mount Everest is known in Nepali as "Sagarmāthā" and in Tibetan as "Chomolungma");

5.10.9.   use different categories for the geographic kind within which a specific entity is classified (e.g., the geographic kind "river" has two translations in French: "fleuve" when a river flows to the sea, "rivière" in any other circumstance).

5.11   Rivers, *Fleuves,* and *Revières*

Considering conventional objects, we have already said that geographic categorizations can be marked by differences in the ways different languages slice the world (see Sect. 5.9). "Terms like "strait" and "river" represent, in this sense, arbitrary partitions of the world of water bodies. The English language might have evolved with just one term, or three terms, comprehending the range of phenomena stretching between strait and river or, in French, between "détroit" and "fleuve"" (Smith and Mark 1998, p. 317).

Different languages might also contain different categories for the classification of the geographical entities. Taking the example of Sect. 5.10, the geographical kind "river" has indeed two translations in French: "fleuve" when a river flows to the sea, "rivière" for all the other rivers. "Notice that there is a topological relation between "fleuve" and sea, and between "fleuve" and "rivière", whereas "river" does not bear this kind of relation" (Laurini 2017, p. 62). Therefore, the Tiber might belong to two different categories, namely "river" and "fleuve", according to whether the language of the person who categorizes such an entity is, respectively, English or French. By using the same example, another issue may emerge if a French speaker sees a natural flowing watercourse not knowing its topological relations: in this case, is he seeing a fleuve or a rivière? Maybe, we should also point out that any categorization, in general, seems to require a good knowledge of the domain we want to categorize.

Moreover, in light of these considerations, we need to specify whether different languages require different classifications—since concepts can be different or differently organized. If they do, the challenge concerns how to match such different classifications of the same (geographical) domain. On the contrary, if they do not, we should select a language for our classification, consider the possibility of integrating translations in different languages, and try not to lose the conceptual richness emerging from different languages. For example, a classification of water bodies in English will lose the topological relations expressed by the (French) dichotomy between "fleuve" and "rivière".

5.12   Danube, *Donau,* and Дунав

If the considerations expressed in Sect. 5.11 are generally focused on common names, we should now consider that some geographical entities have proper ones. In the realm of physical geography, only a few points have proper names, such as the North and the South Poles and some mountain summits, a few lines, such as the Equator, the Tropic of Cancer, the Tropic of Capricorn, the Greenwich Meridian, and the Polar Circle and some solids such as lakes, seas, oceans, and so forth. Conversely, within the human geography the list of proper names is so long that we may look at the discipline as the realm of geographical proper names. Such names might give rise to a number of conundrums on toponymy, which can also be interesting for the debate on geographical entities. According to Laurini (2017), the conundrums might regard:

5.12.1.   homonymy—the fact that a proper name can be the name of two (or more) different geographical entities (i.e., "Mississippi" is the name of a river and of a state);

5.12.2.   endonym/toponym—the former is the local name in the official language of the country or in a well-established language occurring in that area where the feature is located (i.e., "Venezia" in Italian). However, potentially all geographical entities may also have different names (several toponyms) in countries with different official languages (i.e., "Brussel" in Flemish, "Bruxelles" in French);

5.12.3.   exonym, which is a name in languages other than the official one (i.e., "Venice" in English or "Venice" in French);

5.12.4.   archeonym, which is a name that existed in the past (i.e., "Byzantium" and "Constantinople" for Istanbul);

5.12.5.   abbreviations ("L.A". for Los Angeles) and nicknames ("Big Apple" for New York);

5.12.6.   place with multiple names (i.e., in New York City, "Sixth Avenue" is also known as "Avenue of the Americas");

5.12.7.   variations about the way to write some names (i.e., "3rd Street", "Third Street", "Third St");

5.12.8.   transcriptions, for example "Peking" became "Beijing" after a change
           of transcription to the Roman alphabet, but the capital of China has not
           modified its name in Chinese.

If Laurini's list (and examples) is still not enough, we might add, for example, the
case of the river "Danube" that assumes different names in the different countries it
crosses: "Donau" in Germany and Austria, "Dunaj" in Slovakia, "Duna" in Hungary,
"Dunav" in Croatia and Serbia, "Dunav" and "Дунав" in Bulgaria, "Dunărea" in
Romania and in Moldova, and "Dunaj" and "Дунай" in Ukraine. So, what is the
proper name of the river? May we assign a name to that river as a whole? If no, is
the name of such a river composed by the sum of the names given by the ten nations
that it flows through? Or again, which is the name of the river when it separates
two different nations? Should we maybe think that the river is composed of different
parts, each of which has a different proper name? More in general, might the case
of the Danube (as well as the previous examples) involve some sort of vagueness in
the (linguistic) referent and/or in the entity/ies in question?

5.13 Vagueness

Considering the vagueness not only as a pervasive phenomenon of human thought
but also of the geographical world is up for discussion in the current geo-ontological
debate.[11] According to Varzi (2001), virtually every geographic word and concept
suffers from it, and questions such as "How small can a town be?", "Where does a
hill begin?", "How long must a river be?", and "How many islands does it take to
have an archipelago?" are perfectly legitimate. Moreover, vagueness is not exclusive
to common names: the name "Everest", for example, is just as vague as mountains,
hills, towns, and so forth can be, giving rise to its own kind of soritical paradox.

In the same article, the author distinguishes two different kinds of vagueness: *de
re* and *de dicto.*

In line with the former, the vagueness exhibited by geographic names and descriptions should be conceived as ontological, and not as purely epistemic. Accordingly,
"a vague term is one that refers to a vague object, an object the spatial or temporal
boundaries of which are genuinely 'fuzzy'". Therefore, the name "Everest" is vague
insofar as the entity Everest is vague:

> there is no objective, determinate fact of the matter about whether the borderline hunks are
> inside or outside the mountain called "Everest". The same applies to deserts, lakes, islands,
> rivers, forests, bays, streets, neighborhoods, and many other sorts of geographic entities. On
> the *de re* reading, these entities have vague names because they are genuinely vague denizens
> of reality (Varzi 2001, p. 52)

Conversely, the *de dicto* (semantic) reading maintains that geographic vagueness
"lies in the representation system (our language, our conceptual apparatus) and not

---

[11]See Mandelbrot (1967), Sarjakoski (1996), McGee (1997), Bennett (2001), Varzi (2001).

in the represented entity". In other words, "to say that the referent of a geographic term is not sharply demarcated is to say that the term vaguely designates an object, not that it designates a vague object". Accordingly, there is no such thing as a vague mountain. Rather, there are many things where we conceive a mountain to be, each with its precise boundary, and when we say "Everest" we are just being vague as to which thing we are referring to. That is to say that there are several different "ways of tracing the geographic limits of Mount Everest, all perfectly compatible with the way the name is used in ordinary circumstances". In the end, each one of a large variety of slightly distinct aggregates of molecules has an equal claim to being the referent of the vague name "Everest". But each such thing is precisely determinate (Varzi 2001, pp. 54–55).

5.14   (Geographical) Kinds and Properties

In this chapter, the term "entity" has been generally used as synonym of "object" for indicating, in the realm of geography, something like regions, parcels of land, waterbodies, roads, buildings, bridges, and so on, as well as parts and aggregates of all these things. However, the association between entity and object risks being too restrictive for the description of the geographic domain insofar as such a domain may also comprehend other sorts of entities like kinds, properties, relations, boundaries, events, processes, facts, qualities, and so forth.

Geographical kinds, for example, tell us under which category an object falls. In other words, what an object is. For instance, if we consider the three sentences:

5.14.1.   the Nile is a river;
5.14.2.   Bucharest is a city;
5.14.3.   Everest is a mountain.

The terms "river", "city", and "mountain" represent three (possible) examples of geographical kinds that have objects, respectively, "Nile", "Bucharest", and "Everest", as their instances. Generally speaking, Rosch has proposed that (natural) kinds are seen as possessing a radial structure, having prototypes of more central or typical members surrounded by a penumbra of less central or less typical instances (Rosch 1973, 1978). In the geographical domain, Casati et al. (1998) have also emphasized that the entities to which geographers refer are of a different kind and can be distinguished in two main categories, corresponding to the traditional distinction between physical and human geography. On the one hand, there are mountains, rivers, deserts, and so on. On the other hand, there are socioeconomic units: nations, cities, real estate subdivisions—the spatial shadows cast by different sorts of systematically organized human activity. The correspondence between (these two) branches of geography (human and physical) and different sorts of geographical kinds seems to support the idea that, in principle, each different branch (and sub-branch) of geography might be characterized by specific sorts of geographical kinds.

Finally, we should also consider that geographical kinds and objects might also be characterized by (geographical) properties, that is, entities which can be predicated on objects and kinds, or attributed to them (Orilia and Swoyer 2017). Examples of geographical properties may be "has a population of", "has a catchment area of", and so forth. In addition to expressing what things are said to bear, possess, or exemplify, properties help us in the categorization of different geographical kinds and objects. For instance, the property "elevation" (as well as "volume", "relief", etc.) might help us in categorizing a landform as a mountain, a hill, and so forth.

5.15 Relations, Fields, and Time

In addition to kinds and properties, our list of geographical entities can also comprehend items such as relations, which, in turn, might be divided into the following:

5.15.1. mereological, topological, and spatial relations;
5.15.2. (as well as) different sorts of mixed cases of relations among geographical objects.[12]

How may we consider a relation as properly geographic then? Is there a difference, for example, between spatial and geographic relations? According to Laurini, in spite of such a difference not being so clear-cut, "we can say that spatial relations are seen more abstract whereas geographic relations are grounded in the Earth": that is, geographic relations link two or more objects located on the Earth (Laurini 2017, p. 83). Obviously, this does not mean that spatial relations are not also commonly used in the geographical domain, even if they can also be used in other domains such as robotics and medical imagery. Examples of geographical relations can be the relation "is north/south of", as well as specific connections between geographical objects such as rivers ("is a tributary of"), roads ("crosses"), city and country ("is the capital of"), and so forth.

Geographical continuous fields represent another ontological conundrum in the domain of geographical entities. Indeed, on the one hand, we have the position of Smith and Mark (1998), according to which an adequate ontology of geographic kinds should embrace not only categories of *discreta* but also *categories* that arise in the realm of continuous phenomena. On the other hand, Laurini (2017) says that the introduction of a theory of continuous fields might help us in *representing*, especially within geographic information systems (GIS), continuous phenomena such as temperature, pressure, wind, elevation, or air pollution, which can be matters of geographical interest. All this also means to underline that, particularly with GIS, "there is also conceptual interaction with geographical entities that is mediated through mathematical models and through computer representations" (Smith

---

[12]For an analysis of these relations, see Chap. 3.

and Mark 1998, p. 312). Now, if the point concerns whether to consider (continuous) fields as properly geographical *entities* or *tools* involved in the representation of geographical entities, the doubt seems to involve the relation between the geographical reality and the tools that help to describe it. In other words, should we also include such tools or, more in general, entities coming from the domain of geographic representation in our list of geographical entities?

Finally, we should also spare a few words on the dimension of time so as to avoid considering the geographic reality in a static perspective. This means not only contemplating the diachronic and synchronic identity of geographic entities, but also, according to Egenhofer and Mark (1995), regarding geographic space and time as tightly coupled. For instance:

> Many cultures have pre-metric units of area that are based on effort over time (Kula 1983). The English *acre* (Jones 1963; Zupko 1968, 1977), the German *morgen* (Kennelly 1928), and the French *arpent* (Zupko 1978) all are based on the amount of land that a person with a yoke of oxen or a horse can plow in one day or one morning. There have been similar measures for distance, such as how far a person can walk in an hour, or how far an army can march in a day (Egenhofer and Mark 1995, p. 7).

5.16 Boundaries

As stated in Sect. 5.8, one of the many approaches to identifying geographical entities starts with the specification of their boundaries. But what are geographical boundaries? What is their relation to the entities they demarcate? Is it mereological? Might boundaries exist also without the entities they separate? Should we include them in our list of geographical entities?

Without claiming to be exhaustive, we can say that the geo-ontological debate[13] has generally distinguished between two main sorts of theories on (geographical) boundaries: realist and eliminativist theories (Varzi 2015). Realist theories consider boundaries as lower-dimensional entities: boundaries are ontological parasites, which cannot be separated and exist in isolation from the entities they bound. Realist theories may differ significantly, however, with regard to how such dependent, lower-dimensional entities relate to the extended entities they bound. With reference to the boundary between Maryland and Pennsylvania, Varzi has distinguished four main views of such theories:

5.16.1. the first view maintains that the boundary may belong neither to Maryland nor to Pennsylvania;

5.16.2. according to the second one, the boundary must belong either to Maryland or to Pennsylvania, though it may be indeterminate to which of the two states it belongs;

---

[13] See for example Mark and Csillag (1989), Smith (1995), Burrough and Frank (1996), Zimmerman (1996), Smith and Varzi (1997, 2000), Casati et al. (1998), Smith and Mark (1998), Casati and Varzi (1999), Varzi (2007), Russell (2008), Varzi (2016).

5.16.3. the third says that the boundary may belong both to Maryland and to Pennsylvania, "but the relevant overlap is *sui generis* precisely insofar as it involves lower-dimensional parts. Boundaries do not *take up* space and so, on this theory, it is not implausible to say that, for example, the Mason–Dixon line belongs to both Maryland and Pennsylvania";

5.16.4. the last one maintains that there really may be two boundaries, one belonging to Maryland and one belonging to Pennsylvania, "and these two boundaries would be co-located—that is, they would coincide spatially without overlapping mereologically" (Varzi 2015).

Conversely, eliminativist theories move from the idea that talking of boundaries involves some sort of abstraction. Among such theories, substantivalists about space–time "see the abstraction as stemming from the relationship between a particular and its spatiotemporal receptacle, relying on the topology of space–time to account for our boundary talk when it comes to specific cases". If one is not a substantivalist about space and/or time, she/he can describe the abstraction as invoking the idea of ever thinner layers of the bounded entity. On this account, "boundary elements are not included among the primary entities, which only comprise extended bodies, but they are nonetheless retrieved as higher-order entities, viz. as equivalence classes of convergent series of nested bodies" (Varzi 2015).

5.17 On Non-Existent and Abstract Geographical Entities

On Tuesday July 5, 1955, the Australian newspaper *The Age* wrote that the Philippine Air Force was searching the South China Sea for a mysterious island settlement called the "Kingdom of Humanity". The reason for this mission was that the Philippine President wanted to know whether such a place actually existed. If it had been, the Philippine President wanted to determine whether it was a legitimate settlement within the territorial Philippines (Middleton 2015). But, if it were not the case, could we have included the "Kingdom of Humanity" within the list of geographical entities? In other words, does the notion of existence determine what we can legitimately consider as a geographical entity?

Another example may be represented by Lunezia, a geographical region that is meant to include the Italian provinces of La Spezia, Massa-Carrara, Parma, Piacenza, Reggio nell'Emilia, Mantua, and part of the territories of Cremona and Lucca. Since 1946, the debate on the possible constitution of Lunezia has not (yet) led to the institution of such a geographical region. And yet, does the (ongoing) debate legitimize the inclusion of Lunezia within the geographical entities? Or does the fact that Lunezia never had a spatiotemporal existence on the Earth exclude such an entity from the realm of geographical entities?

If all these examples are not enough, let us go back to the Era of Partition, when Poland did not have any territory to call its own. Now, should we include Poland among the geographical entities also in that era? More precisely, if we wanted to carry out an inventory of geographical entities of that period, does the fact the Poland did not have a territory (or a spatiotemporal existence during such an era) allow us to exclude Poland from that inventory? If no, we could also include, within the list of geographical entities, entities that did not have (and maybe that will no longer have) a spatial location, such as the Holy Roman Empire or the Maritime Republics, and (maybe) entities such as Kosovo, Timor Est, and South Sudan, which had not (yet) had a spatiotemporal existence during that period of time. If yes, we should perhaps justify how, for example, a non-geographical entity can give the right to (re)claim a territory as its own land, such as Poland after the Era of Partition or nowadays with Kurdistan.

5.18 Historical Entities

Until McCarthy completed his work, Siamese provinces were not geographically well-described. A province existed in a particular place but the place did not define it. The land itself was almost coincidental. What mattered were the people. And where a boundary did exist, it was seldom a continuous line. It wasn't even a zone. In fact it only occurred where it was needed, such as along a track or pass used by travellers. In other places, where people seldom set foot, there was no point in deciding a boundary. Further, borders between adjacent kingdoms did not necessarily touch, often leaving large unclaimed regions of forest, jungle or mountains. And in practice it was quite possible for towns to have multiple hierarchical relations of authority with more than one ruler and hence – disturbingly for Mr McCarthy – to be part of more than one state (Middleton, 2015).

In Sect. 5.9, it has been said that different cultural frameworks (as well as different languages and beliefs) may describe the same geographical reality in diverse ways, in terms of categorizations, entities, boundaries, and so forth. This means that cultural environment plays a fundamental role in determining our list of geographical entities. However, we should also remark that such a cultural framework does not change only on the basis of the geographical context. Indeed, the advancements of geography as a discipline and the historical context can also have a strong influence on it.

About the influence of the historical context—besides the case of Siamese provinces provided by Middleton—we can, for example, consider if there is a difference between contemporary (military) encampments and Roman *Castra* (or *Hiberna*)—regardless of whether or not Castra had become cities. So, should we include such entities in our geographical inventory? Do contemporary military encampments and Roman *Castra* represent the same geographical entity? Another issue might arise from territories occupied by nomadic populations, which could change according to seasons, food resources, and so forth. In this case, we could ask, is there a geographical entity defined by the territory occupied by a population in a specific period of time, even if that population did not have an ongoing territory to call its own? If yes, can it be an entity that describes the ancient world but not the contemporary one? More generally, do we use the same geographical concepts that, for example, the Greeks and/or Romans used? Did the notion of boundaries have the same meaning that it has today? Did, for example, the term "Gaul" denote a crisp region with clear-cut boundaries or rather the territory occupied by Celtic tribes with *de re* vague boundaries?

To conclude, we should consider mythological places such as Atlantis, Biringan City, Cloud cuckoo land, Paititi, and Mu. Are they geographical entities, at least for some cultures in certain periods of time? If yes, should we include them in our list of geographical entities? Just to add further hurdles, we might also consider the case of Thule and the several theories about its possible location, which include, among others, the coastline of Norway, Iceland, Greenland, Orkney, Shetland, Faroe Islands and Saaremaa. Obviously, if we imagine a map that shows all these locations, then we would be hardly inclined to consider Thule as the mereological sum of all the locations ascribed to it. At the same time, it would be unlikely to consider the various Thules represented on the map (with different conditions of identity) as duplicates of the same geographical entity. Perhaps, we could take the various points that locate Thule on that map as indicating different geographical entities, to which different authors have attributed the same connotation. Perhaps, we could also consider the possibility of geographical entities with multiple locations.[14]

5.19 Complex Geographical Entities

Generally speaking, geographic objects are complex entities: that is, they have proper parts and/or components. Moreover, geographic objects can be connected or contiguous, but they can also be scattered or separated. Sometimes they are closed (e.g., lakes), and some others are open (e.g., bays). Note that the above concepts of contiguity and closure are topological notions, and thus an adequate ontology of geographic objects must contain also a topology—a theory of boundaries and interiors, of connectedness and separation—that is integrated with a theory of parts and wholes, or mereology (Smith 1996).

To say that some geographical entities may be complex means that such entities are made up of other geographical entities: for example, a nation can be divided into regions, provinces, and so forth, a city can contain geographical entities such as buildings, streets, and so on. They can all be seen either from a mereological approach (part/whole relations), or from a topological point of view (contain relation). However, we should remark that a geographical entity might also have components which are not strictly geographical. To put it clearly, if a geographical entity such as a forest might be defined as a large area covered chiefly with trees and undergrowth, may we consider these trees, their leaves, roots, and atoms as geographical entities? Moreover, a geographical entity such as a forest might also have (arbitrary) spatial parts: for example, the north side of the forest and the south one. But then, should we include such spatial parts within our list of geographical entities?

---

[14]See also Chap. 3.

Another point to mention is that the hierarchical structure exhibited by, for example, the relations between a nation with its regions, administrative subdivisions, and so forth is not the only possible structure that geographical entities might show. Indeed, according to Laurini, as there are different kinds of roads, turnpikes, streets, etc., seldom, in this case, can a sort of hierarchy be defined? Moreover, some geographical entities can contain specific parts of other geographical entities. In other words, a geographical entity may, in principle, belong to two or more different geographical entities, making it difficult to think about a hierarchical structure. For example, Via Emilia (SS 9) crosses different Italian provinces such as, among others, Rimini, Bologna, Reggio nell'Emilia, and Parma. Furthermore, a geographical entity may also belong to two or more different hierarchies that, for instance, describe different branches of geography (as a discipline). In this sense, hierarchies can also presuppose overlaps. For example, Lake Iseo can be seen as an instance of the class "Lakes" that, in turn, is a subclass of the class "Water Bodies" (physical geography). At the same time, Lake Iseo can be considered as belonging to the region Lombardy that in turn is a proper part of the nation Italy and so forth (political geography). However, we should not forget that the presence of a hierarchy does not exclude eventual relations among classes at the same level or belonging to different branches of the same hierarchy (Bittner and Smith 2008).

5.20 Hierarchical Structures

To talk about (geographical) hierarchies, it may be useful to introduce the meaning of two different terms that I use in this Sect.: "hyperonym" and "hyponym". The two terms are the (opposite) names of places with a hierarchy: for instance, Europe is a hyperonym of Italy, whereas Italy is a hyponym of Europe (Laurini 2017). In contrast, a "meronym" may be considered as a name of a part of a place without a hierarchy: for instance, the Adriatic Sea is a meronym of the Mediterranean Sea.

Now, could we benefit from thinking in terms of hierarchy in distinguishing between what is geographical and what is not? Perhaps we should first consider whether or not the hierarchy can be inclusive for all the geographical entities on our list. Accordingly, the point might be to circumscribe such a hierarchy, starting from the top hyperonym and lowest hyponyms.

About the top hyperonym, a fundamental question might be: is there something geographic to which anything uncontentiously belongs? Semantically speaking—given that the term geography comes from the Greek words "gê" ("Earth") and "graphein" ("to write, draw") and thus it means "to write and draw about the Earth"—a possible answer can be the Earth: every geographical entity belongs to the all-inclusive geographical entity "Earth". Now, if such an answer may have some supporters, we should nonetheless pay attention to, at least, two different issues. The first one is to keep geography from collapsing into its cartographical dimension (or better, to not reduce geography to cartography). The second issue, strictly related to the first one, concerns the fact that geography is also devoted to the study of human activities, cultures, economies, interaction with the environment and relations with and across space and place. Of course, such human dynamics can have effects on the Earth, by producing something that can be analyzed through a study of the Earth. However, we can also assume that, even though human dynamics might have an impact on the Earth, they are something more. Accordingly, the "Earth" does not complete the entire domain of geography.

Now, what about the lowest hyponyms? An idea might be to consider only those geographical entities that are not complex. Consequently, the lowest geographical hyponym (LGH) is a geographical entity that does not contain (or that is not composed of) other geographical entities. (Obviously, that does not mean that a LGH cannot contain other entities, which, in turn, should not be geographic.) However, without a definition of geographical entities, a clear-cut identification of a LGH might be difficult. For example, if considering a street as a geographical entity seems to be uncontentious, might we say the same also for shoulders, (emergency, cycle) lanes, roadways of that street? What is/are the LGH(s) in this context? And what if we consider the relation between ponds and lakes? Are ponds hyponym of lakes or are lakes and ponds both categories at the basic level, mainly distinguished by size? What is/are the LGH(s) among a forest and its north and south sides? Finally, we should also consider that LGHs can change according to the different branches of geography we investigate—and consequently every branch of geography might have, in principle, a proper list of geographical entities. For instance, shoulders, roadways, and so forth can be potential examples of LGHs for transportation geography but not for health geography; entities such as airports and tracks may be considered as geographic for some branches of human geography but not for classical geography (Luckermann 1961) and so on.

5.21   Three Thin Red Lines

The aim of this chapter has been to provide a (non-exhaustive) sketch of possible approaches, response attempts, conundrums, and issues arising from the question "What is a geographical entity?". Trying to answer this question is made particularly difficult by the multiplicity of aspects that might influence our answer and that defies a clear-cut systematization. Without claiming completeness, we might summarize such aspects as follows.

The first one emerges from the fact that we can use (many) different conceptualizations for describing geographic space. On the one hand, according to Egenhofer and Mark (1995), such conceptualizations of geographic space may:

5.21.1.   reflect the differences between perceptual and cognitive space (Couclelis and Gale 1986);

5.21.2.   be based on different geometrical properties, such as continuous vs. discrete (Egenhofer and Herring 1991; Frank and Mark 1991);

5.21.3.   depending on scale or difference in the types of operations, we would
           typically employ in everyday life and/or in scientific reasoning (Zubin
           1989).

On the other hand, as I have often remarked, different conceptualizations of
geographical space can also emerge from the ways in which different languages and
cultures—as well as the various geographical branches and perspectives—structure
and systematize the world itself (Oakes and Price 2008). In this sense, as Smith and
Mark (2001) suggest, work involving formal comparisons of geospatial and cartographic data standards and dictionary definitions in a variety of languages might
also provide an important starting point for combining quantitative, i.e., measurable
geographic phenomena described by different scientific disciplines, with qualitative
geographical descriptions of reality also emerging from areas of human-geographical
reasoning.

The second factor is that sometimes we may have some difficulties in distinguishing the domain of the real world from the domain of computational and mathematical representations, and both from the cognitive domain of reasoning, language
and human action (Smith and Mark 1998). Moreover, it might sometimes be difficult
to provide a clear-cut distinction between the real world and the tools that we can use
to describe it (Laurini 2017). For example, should we consider a compass as a tool
capable of describing part of the geographical world or also as a proper geographical
entity? And what about items such as GPS coordinates, longitude, latitude, and so
forth (Crampton 2010)? Do mathematical entities exist in the geographical world?
And geometric ones? And geographical entities which are derived from technology?
Can GIS enrich our geographical inventory with new kinds of geographical entities?

The third factor concerns geography itself and specifically its development (and/or
advancement), which does not only affect geography as a discipline but also the world
that it describes. Take for example modifications of boundaries, the formulation of
(the notion of) nation states, the presence of airports on our current maps or, again, the
possibilities given by augmented reality for geography. Take also the introduction of
new scholarly fields in geography such as night studies (Gwiazdzinski and Chausson
2015) and border studies (Newman 2006; Kolossov and Scott 2013), or the birth of
intellectual movements or paradigm shifts such as the spatial turn (Warf and Arias
2009). Take finally any example that represents a potential change in some parts
of contemporary geography (Gomes and Jones 2010) if compared, for instance, to
classical geography (Lukermann 1961; Bianchetti 2008), in terms of assumptions,
tools, methods of investigation and domain to describe. Now, may we assume that (at
least) some of these changes, developments and/or advancements, which introduced
new ways of slicing, shaping and interpreting the geographical world, could/can/will
create new perspectives for distinguishing between what is geographical and what is
not?

5.22   From Multiple (Ways of Doing) Geographies to Multiple (Kinds of) Geographical Entities

If the answer to the above question is yes, as I presume, providing an exhaustive definition of geographical entity (as well as a full list of them) would be made even more difficult, to the point of running the risk of being too restrictive for what a geographical entity could be in the past and will be in the future. For that, although not offering a definition can hardly seem very precise in distinguishing what is geographical from what is not, I think that the possible imprecision of such a definition would be even worse. The issue, in this case, would be to hinder the process of theory construction, especially for what concerns how best to interpret new (possible) geographical evidence. In other words, the idea is that since geographers (as well as GIS scientists and geo-ontologists) approach the task of theory construction under the guidance of some ontological assumptions, the greatest contributions of analyzing the notion of geographical entity would be essentially two. The first one is simply to chart the possibilities of existence (Lowe 1989, 2006). The second contribution is providing us with the conceptual tools wherewith to categorize the world's contents in view of the heterogeneity of the geographical debate, trying to keep open minds as to how we might interpret new geographical aims, perspectives, and points of view.

Accordingly, the idea behind this chapter is that of subordinating every normative claim on the notion of geographical entity to (as well as of enriching our descriptive approaches with) the factors we underlined, which can be summarized as follows: (1) There exist multiple conceptualizations of the geographical world. (2) Different languages and cultures slice such a world in different ways. (3) The geographical world has changed and will change over time. (4) Geography (as a discipline) has changed and will change over time too, by modifying its perspective, tools, domains of investigation and aims. Consequently, what was, is, and will be considered as non-geographic could be considered as geographic, and vice versa. (5) There were, are, and will be different kinds of geographies as well as different geographical branches, each of which had, has, and will have different tools, aims, and vocabularies. (6) The introduction of new scholarly fields and new technologies, the birth of intellectual movements or paradigm shifts can/will influence geography as a discipline.

This means that there are multiple, alternative, and overlapping views on geographical reality, and the same reality can be represented and sliced in different ways. Accordingly, the aim of an investigation on the notion of geographical entity should be to provide some platforms for integrating such alternative views. Its task is thus practical in nature and is subject to the same practical constraints experienced in all scientific activity. Consequently, such a geo-ontological investigation will always be a partial and imperfect edifice subject to correction and enhancement, so as to meet new scientific needs (Smith and Klagges 2008).

References

Bateman J, Farrar S (2004) Towards a generic foundation for spatial ontology. In: Varzi AC, Vieu L (eds) Formal ontology in information systems. IOS Press, Amsterdam, pp 237–248

Bennett B (2001) What is a forest? On the vagueness of certain geographic concept. Topoi 20(2):189–201

Bianchetti S (2008) Geografia storica del mondo antico. Monduzzi, Bologna

Bishr Y (2007) Overcoming the semantic and other barriers to GIS interoperability: seven years on. In: Fischer P (ed) Classics from IJGIS. Twenty years of the international journal of geographical information science and systems. CRC Press, Boca-Raton-London-New York

Bittner T, Smith B (2008) A theory of granular partitions. In: Munn K, Smith B (eds) Applied ontology. An introduction. Ontos-Verlag, Berlin

Bonnett A (2008) What is geography? Sage, London

Burrough PA, Frank AU (eds) (1996) Geographic objects with indeterminate boundaries. Taylor & Francis, London

Casati R, Smith B, Varzi AC (1998) Ontological tools for geographic representation. In: Guarino N (ed) Formal ontology in information systems. IOS Press, Amsterdam, pp 77–85

Casati R, Varzi AC (1999) Parts and places. MIT Press, Cambridge

Casti E (2015) Reflexive cartography. A new perspective on mapping. Elsevier, Amsterdam, Oxford, Waltham

Crampton JW (2010) Mapping. A critical introduction to cartography and GIS. Wiley-Blackwell, Oxford, New York

Couclelis H (1998) Space, time, geography. Geogr Inf Syst 1:29–38

Couclelis H, Gale N (1986) Space and spaces. Geogr Ann 68(B):1–12

Egenhofer M, Herring J (1991) High-level spatial data structures for GIS. In: Maguire D, Goodchild M, Rhind D (eds) Geographical information systems, vol 1. Principles. Longman, London, pp 147–163

Egenhofer M, Mark DM (1995) Naive geography. In: Frank AU, Kuhn W (eds) Spatial information theory: a theoretical basis for GIS. In: Proceedings of the second international conference. Springer, Berlin, Heidelberg, pp 1–15

Elden S (2009) Philosophy and human geography. In: Kitchen R, Thrift N (eds) International encyclopedia of human geography. Elsevier, Oxford, pp 145–150

Elford W (2012) A multi-ontology view of ergonomics: applying the cynefin framework to improve theory and practice. Work 41(1):812–817

Epstein B (2017) What are social groups? Their metaphysics and how to classify them. Synthese. https://doi.org/10.1007/s11229-017-1387-y

Farinelli F (2003) Geografia. Un'introduzione ai modelli del mondo. Einaudi, Torino

Farinelli F (2009) La crisi della ragione cartografica. Einaudi, Torino

Frank A, Mark DM (1991) Language issues for GIS. In: Maguire DJ, Goodchild MF, Rhind DW (eds) Geographical information systems. Principles and applications, vol 1. Longman, London, pp 147–163

Galton A (2003) On the ontological status of geographical boundaries. In: Duckham M, Goodchild MG, Worboys MF (eds) Foundation of geographic information science. Taylor & Francis, London, New York, pp 151–171

Gomez B, Jones JP III (eds) (2010) Research methods in geography: a critical introduction. Wiley-Blackwell, West Sussex

Grenon P, Smith B (2007) Persistence and ontological pluralism. In: Kanzian C (ed) Persistence. Springer, New York, pp 33–48

Günzel S (2001) Geophilosophie. Nietzsches philosophische Geographie. Akademie, Berlin

Gwiazdzinski L, Chausson N (eds) (2015) Urban nights. J Urban Res 11 (Special issue). https://journals.openedition.org/articulo/2595

Inkpen R, Wilson G (2013) Science, philosophy and physical geography. Routledge, London

Jones S (1963) Weights and measures: an informal guide. Public Affairs Press, Washington, D.C.

Kennelly A (1928) Vestiges of pre-metric weights and measures persisting inmetric-system Europe, 1926–1927. The Macmillan Company, New York

Kolossov V, Scott J (2013) Selected conceptual issues in border studies. Belgeo 1. http://belgeo.rev ues.org/10532

Kula W(1983) Les Mesures et Les Hommes. Paris: Maison des Sciences de L'Homme [Translated from Polish by J Ritt; Polish edition 1970]

Laurini R (2017) Geographic knowledge infrastructure: applications to territorial intelligence and smart cities. ISTE-Elsevier, London

Lowe EJ (1989) Kinds of being: a study of individuation, identity and the logic of sortal terms. Basil Blackwell, Oxford, New York

Lowe EJ (2006) The four-category ontology: a metaphysical foundation for natural science. Clarendon Press, Oxford

Luckermann F (1961) The concept of location in classical geography. Ann Assoc Am Geogr 51(2):194–210

Mandelbrot B (1967) How long is the coast of Britain? statistical self-similarity and fractional dimension. Science 156:636–638

Mark DM, Csillag F (1989) The nature of boundaries on 'area-class' maps. Cartographica 26:65–78

McGee V (1997) Kilimanjaro. Can J Philos 23:141–195

Middleton N (2015) An atlas of countries that don't exist. A compendium of fifty unrecognized and largely unnoticed states. MacMillan, London

Newman D (2006) The lines that continue to separate us: borders in our 'borderless' world. Prog Hum Geogr 30(2):143–161

Oakes TS, Price PL (eds) (2008) The cultural geographer reader. Routledge, New York

Orilia F, Swoyer C (2017) Properties. In: Zalta EN (ed) The Stanford encyclopedia of philosophy. https://plato.stanford.edu/archives/win2017/entries/properties/

Pattinson WD (1963) The four traditions of geography. J Geogr 63(5):211–216

Rosch E (1973) On the internal structure of perceptual and semantic categories. In: Moore TE (ed) Cognitive development and the acquisition of language. Academic Press, New York, pp 111–144

Rosch E (1978) Principles of categorization. In: Rosch E, Lloyd BB (eds) Cognition and categorization. Erlbaum, Hillsdale, pp 28–49

Russell JT (2008) The structure of gunk: adventures in the ontology of space. Oxf Stud Metaphys 4:248–274

Sala M (2009) Geography. In: Sala M (ed) Geography. EOLSS Publisher, Oxford, Encyclopedia of life support systems, pp 1–56

Sarjakoski T (1996) How many lakes, islands and rivers are there in Finland? A case study of fuzziness in the extent and identity of geographic objects. In: Burrough PA, Frank AU (eds) Geographic objects with—indeterminate—boundaries. Taylor & Francis, London, pp 299–312

Smith B (1994) Fiat objects. In: Guarino N, Pribbenow S, Vieu L (eds) Parts and wholes: conceptual part-whole relations and formal mereology. In: Proceedings of the ECAI94 workshop. Amsterdam, ECCAI, pp 15–23

Smith B (1995, September 21–23) On drawing lines on a map. In: Frank A, Kuhn W (eds) Spatial information theory—a theoretical basis for GIS. In: Proceedings, international conference Cosit'95, Semmering, Austria. Lecture notes in computer science, vol 988. Springer, Berlin, pp 475–484

Smith B (1996) Mereotopology: a theory of parts and boundaries. Data Knowl Eng 20:287–303

Smith B, Klagges B (2008) Bioinformatics and philosophy. In: Munn K, Smith B (eds) Applied ontology. An introduction. Ontos-Verlag, Berlin

Smith B, Mark DM (1998) Ontology and geographic kinds. In: Poiker TK, Chrisman N (eds) Proceedings of the eighth international symposium on spatial data handling (Burnaby, British Columbia, International Geographical Union), pp. 308–320

Smith B, Mark DM (2001) Geographical categories: an ontological investigation. Int J Geogr Inf Sci 15(7):591–612

Smith B, Varzi AC (1997) Fiat and bona fide boundaries: towards an ontology of spatially extended objects. In: Hirtle SC, Frank AU (eds) Spatial information theory: a theoretical basis for GIS. Lecture notes in computer science, vol 1329. Springer, Berlin, pp 103–119

Smith B, Varzi AC (2000) Fiat and bona fide boundaries. Res 60(2):401–420

Tanca M (2012) Geografia e filosofia. Franco Angeli, Milan

Varzi AC (2001) Vagueness in geography. Philos Geogr 4(1):49–65

Varzi AC (2007) Spatial reasoning and ontology: parts, wholes and location. In: Aiello M, Pratt-Hartmann I, van Benthem J (eds) Handbook of spatial logics. Springer, Berlin, pp 945–1038

Varzi AC (2015) Boundary. In: Zalta EN (ed) The Stanford encyclopedia of philosophy. https://plato.stanford.edu/archives/win2015/entries/boundary/

Varzi AC (2016) On drawing lines across the board. In: Zaibert L (ed) The theory and practice of ontology. Palgrave Macmillian, London, pp 45–78

Warf B, Arias S (eds) (2009) The spatial turn interdisciplinary perspectives. Routledge, London, New York

Westerhoff J (2005) Ontological categories. Clarendon Press, Oxford

Zimmerman DW (1996) Could extended objects be made out of simple parts? Res 56:1–29

Zubin D (1989) Untitled. In: Mark D, Frank A, Egenhofer M, Freundschuh S, McGranaghan M, White RM (eds) Languages of spatial relations: initiative two specialist meeting report. Technical paper 89–2, National Center for Geographic Information and Analysis, Santa Barbara, CA, pp 13–17

Zupko R (1968) A dictionary of english weights and measures. The University of Wisconsin Press, Madison

Zupko R (1977) British weights and measures: a history from antiquity to the seventeenth century. The University of Wisconsin Press, Madison

Zupko R (1978) French weights and measures before the revolution: a dictionary of provincial and local units. Indiana University Press, Bloomington

Part III
The Philosophy of Geo-Ontologies

Chapter 6
Geo-Ontologies: From the Spatial Turn to Geographical Taxonomy

Abstract While by discussing notions of spatial representation, boundaries, and geographical entities Chaps. 3–5 have been essentially speculative, this chapter aims to analyze geo-ontologies as an IT/computer application of the theoretical investigation presented so far. Section 6.1 explores the emergence of geo-ontologies from the spatial turn and sketches their general as well as specific goals. Sections 6.2 and 6.3 are, respectively, dedicated to show the most relevant examples of taxonomies derived from the IT domain and to underline the absence of a classification suitable for spreading geo-ontologies in the geographical debate. Finally, Sect. 6.4 is concerned with outlining a taxonomy grounded on the distinction between spatial, physical, and human geo-ontologies. The idea behind this taxonomy is to relate geo-ontologies to the geographical debate which, in turn, could improve the conceptualizations of such ontologies.

Keywords Geo-ontologies · Human geo-ontologies · Physical geo-ontologies · Spatial geo-ontologies · Spatial turn · Taxonomies

6.1 From the Spatial Turn to the Diffusion of Geo-Ontologies

In talking about geo-ontologies, we should first emphasize that, over the last two decades, geography has undergone a profound conceptual and methodological renaissance, which has transformed it into one of the most dynamic, innovative, and influential areas of studies. This renaissance is specifically defined as "the spatial turn" and involves a reworking of the very notion and significance of spatiality, unfolding across the social sciences and humanities.

Those disciplines have taken up geography in their own way and have become increasingly spatial in their orientation. In particular, social sciences hold that space is a social construction relevant to the understanding of the different histories of human subjects, to the production of cultural phenomena, and to the recognition of geographic dimensions as essential aspects of that cultural development. Moreover, the spatial turn reflects much broader transformations in the economy, politics, and culture of the contemporary world, asserting that:

6.1.1.   we cannot comprehend the production of spatial ideas independently from
          the production of spatiality;
6.1.2.   space is not simply a passive reflection of social and cultural trends, but an
          active participant (Warf and Arias 2009, pp. 1–2).

According to Warf and Arias, five main forces have come together and increased
the relevance of space both at a material and at an ideological level:

6.1.3.   contemporary *globalization*, which refers to a variety of processes that
          does not play out identically in different places, bridges the gap between
          distant cultures, events, and places, entails a comparison among residents
          in many countries, and creates a global scenario that has called attention to
          national differentials. Globalization has undermined commonly held notions
          of Euclidean space by forming linkages among disparate producers and
          consumers intimately connected over vast distances through flows of capital
          and goods. Other factors are involved and strictly connected with globaliza-
          tion, namely migration, tourism, (international) media, offshoring of many
          jobs from the developed world to the developing one, international finance,
          a worldwide space of flows, global deregulation, the decline in transport
          costs, and so forth;
6.1.4.   the rise of *cyberspace* and the *Internet* that allows users to transcend
          distance virtually and to connect effortlessly with others around the world.
          Telecommunication systems have become the central technology of post-
          modern capitalism, vital not only to large and small corporations, but also
          to consumption, personal communication, entertainment, education, poli-
          tics, and numerous other domains of social life. Indeed, cyberspace has
          evolved into an important part of everyday life, dissolving the once-solid
          boundary between the real and the virtual. Moreover, cyberspace has been
          instrumental to the production of complex, fragmented, jumbled spaces of
          postmodernity, all of which have called for mounting scrutiny;
6.1.5.   a profound *shift in social life and (human) identity and subjectivity*, which
          varies from fixed and unified to multiple, alienated, fragmented, and contra-
          dictory. Increasingly, the notion of the autonomous subject standing apart
          from the world he/she observes has been questioned and in its place lies a
          greater pluralistic affirmation of cultural difference based on several axes of
          identity (gender, ethnicity, sexuality, etc.);
6.1.6.   the rapidly rising seriousness of *global ecological and environmental prob-
          lems*. These aspects are now increasingly viewed as approachable only on
          a worldwide basis and inseparable from the global context. Because envi-
          ronmental issues are unevenly distributed across space, and because geog-
          raphy as a discipline has a long history of investigating human–environ-
          ment interactions, space and spatiality have become crucial dimensions in
          understanding and tackling these problems;

6.1.7. the development of *Geographical Information Systems* (GIS). These resources offer new means to analyze spatiality. But GIS is not simply reflective of the new importance of space, but also constitutive of it (Warf and Arias 2009, 5–6).

Such a development of GIS—together with the impulse of the Semantic Web,[1] the innovations in the online cartographic visualization[2] and the demand of systematization, cataloging and mapping of geographic information—has also allowed a proliferation of geo-ontologies[3] in the IT domain.[4]

As we will see in Chap. 7, geo-ontologies can be considered as IT/computer representations of geographical entities, aimed at describing geographical application domains. While their most general aims include accessibility, repeated applicability, informativeness, and completeness, geo-ontologies rarely propose conceptualizations that represent the overall geographical domain; they only consider some specific geographical aspects.[5] Moreover, one single ontology might include elements belonging to different geographical branches, incomplete inventories, vague distinctions, and conceptualizations created by non-professional geographers for whom commonsense plays a central role. All this makes a rigid and unambiguous classification of such ontologies difficult to obtain.

6.2 The Problem of Existing Taxonomies

In the domain of IT/computer sciences, there exist numerous works of classification which structured single (IT/computer) ontologies according to broader classes (Bullinger 2008, p. 172). Without claiming to be exhaustive, the taxonomies of Mizoguchi et al. (1995), Uschold and Grueninger (1996), van Heijst et al. (1996), Guarino (1998), Sowa (2000), and Lassila and McGuinness (2001) represent some of the most relevant examples.

The classification of Mizoguchi et al. (1995) identifies four kinds of ontologies: content ontologies for reusing knowledge, communication ontologies for sharing knowledge, indexing ontologies for case retrieval, and meta-ontologies which are equivalent to what other authors refer to as a knowledge representation ontology.

---

[1] See Berners-Lee et al. (2001).

[2] About this fast-moving field, see Turner (2006), Goodchild (2007), Boll et al. (2008), Hudson-Smith and Crooks (2008).

[3] See Bishr and Kuhn (2000), Câmara et al. (2000), Frank (2001), Kuhn (2001), Rodrìguez and Egenhofer (2004), Visser (2004), Kavouras et al. (2005), Janowicz (2006), Euzenat and Shvaiko (2007), Buccella et al. (2008).

[4] See Battle and Kolas (2012), Perry and Herring (2012), Kyzirakos et al. (2014).

[5] See Chap. 7.

Uschold and Grueninger (1996) published an overview of the three key dimensions along which ontologies can vary: formality (degree of formality of a vocabulary and its meaning), purpose (intended use/application of the ontology), and subject matter (subject of the ontology).

van Heijst et al. (1996) classify ontologies according to two orthogonal dimensions: the amount and type of structure of the conceptualization, and the subject of the conceptualization. Regarding the former dimension, they distinguish among terminological ontologies such as lexicons, information ontologies like database schemata and knowledge modeling ontologies that specify the conceptualizations of knowledge. As for the latter dimension, they identify representation, generic, domain, and application ontologies.

Guarino (1998) orders different types of ontologies according to their level of dependence on a particular task or point of view and distinguishes among top-level, domain, task, and application ontologies.

Sowa (2000) differentiates formal and terminological ontologies by the degree of axiomatization in their definitions. Formal ontologies have their categories and individuals distinguished by axioms and definitions stated in logic or in some computer-oriented language that can automatically be translated into logic. Terminological ontologies do not need to have axioms restricting the use of their concepts.

Finally, Lassila and McGuinness (2001) classify ontologies according to the information the ontology needs to express and to the richness of its internal structure; in particular, they point out the following categories: controlled vocabularies, glossaries, thesauri, informal is-a hierarchies, formal is-a hierarchies, formal instances, frames, value restriction, and general logical constraints (Gómez-Pérez et al. 2004, pp. 26–29).

Despite the fact that the conceptual quality of such taxonomies is not in question, there seem to be reasons to believe that their interest concerns especially the IT/computer characterization of the ontologies. Even by considering only the narrower domain of geo-ontologies, the categorizations proposed by each taxonomy would avoid accounting for any feature which is prominently geographical.

From an IT/geo-ontological perspective, the classification developed by *W3C Geospatial Incubator Group*[6] presents similar problems: it provides an overview and a description of geospatial foundation ontologies with the aim of representing and extending geospatial concepts and properties for use on the World Wide Web. Indeed, the identification of geospatial features, feature types, (geo) spatial relationships, toponyms (place names), coordinate reference systems/spatial reference grids, geospatial metadata, and (geospatial) Web services seems to refer more to specific aspects of geo-ontological analysis rather than to geography as a discipline.

---

[6]See https://www.w3.org/2005/Incubator/geo/XGR-geo-ont-20071023/#ontologies.

6.3   A Geographical Point of View

What is missing, and thus what I want to propose, is a classification suitable for spreading geo-ontologies in the geographical debate. So conceived, such a classification should:

6.3.1. take into account the fundamental distinctions and sub-branches of geography as a discipline (i.e., big-G Geography of Chap. 1);

6.3.2. be integrated with another essential feature of geo-ontologies: the presence of geographical common-sense conceptualizations, denoting a *lower* geography, to be distinguished from *professional* geography (i.e., small-g geography of Chap. 1).

Therefore, the reflection developed so far on applied ontology of geography might be reconsidered to be functional as it makes explicit the assumptions implied by the discipline (see Chaps 3–5) *and* determines how non-professional geographers conceptualize geography (see Chap. 2).

Looking among the handbooks of geography, what constantly recurs is that at least since the late nineteenth century, the conjoint understanding of geography has gradually been augmented by more precise sub-disciplinary pursuits and identities. The most basic of these describes geography as consisting of two fundamental halves: physical and human geography. Generally speaking, physical geography is the science of the Earth's surface, and it is aimed at classifying and analyzing landforms and ecosystems, explaining hydrological, geomorphological and coastal processes, and examining problems such as erosion, pollution, and climatic variability. Instead, human geography usually refers to the study of peoples and geographical interpretations of economies, cultural identities, political territories, and societies. In other words, human geographers analyze population trends, theorize social and cultural changes, interpret geopolitical conflicts, and seek to explain the geography of human economic activities around the world.

> How exactly this division of labor came to be is a most pivotal story of contemporary geography. It is a story about twentieth-century scientific fragmentation, and about different theories on the status of humans vis-à-vis nonhuman nature. [...] Most contemporary academic geographers hold some nominal allegiance to either of geography's "halves" (human or physical), although for important intellectual and philosophical reasons [...] some do resist this division and instead prefer to regard geography as a disciplinary whole, or insist on troubling the conceptual distinction between "the human" and "nonhuman" parts of the world. Some commentators have criticized a perceived widening of the gulf between geography's two halves. Others see human geography as merely a convenient badge for its diverse contents, while still adhering to the principle of a wider, umbrella discipline of geography (including physical geography) (Gibson 2009, p. 218)

Despite some possible criticisms and ambiguities, the distinction captures a central aspect of the geographical reflection, that it has not gone unnoticed even to the philosophical debate as well evidenced by the reflections on geographical entities and boundaries, for which the subdivision between human and physical geography appears quite explicit.[7] However, restricting the classification of the geo-ontologies to this distinction does not allow us to get to a further fundamental feature they share, that is a character at which applied ontology of geography has often shown a specific interest: the spatial representation.[8] Such a feature represents a fundamental issue also for the geographical debate. Pattinson (1963), for example, includes it among the four main traditions of geography, naming it as "spatial analysis". More recently, Sala (2009) splits the whole geographical domain into three different sub-areas, distinguishing among human, physical, and technical geography. About technical geography, the author specifies that the topic presents a range of subjects including geomatics, geodesy, topography, mapping, and some of geography's classical subject matters. Technical geography also introduces the modeling of geographical systems—a field that has been gaining an important place in modern geography.

6.4   A Geo-Ontological Tri-Partition

Sala's tri-partition of (the main) geographical sub-areas might represent an all-embracing classification for geo-ontologies that grasps, specifically, their main contents. By adapting the term "spatial" from Pattinson's taxonomy, my purpose is to distinguish among (see Fig. 6.1):

6.4.1.   spatial geo-ontologies (SGO)[9];
6.4.2.   physical (or natural) geo-ontologies (PGO);
6.4.3.   human geo-ontologies (HGO).[10]

The usefulness of this classification is twofold. First, by being grounded on some essential geographical distinctions, it might de facto introduce geo-ontologies into the geographical debate. Second, it is worth considering potential comprehensiveness: as the tri-partition aspires to be exhaustive for the whole geographical domain (and for its fundamental sub-areas), all geo-ontologies should find a place in the classification. Such comprehensiveness does not coincide with the mutual exclusivity that characterizes the highlighted tri-partition of geography in a broad sense, and for this reason, we should be prepared to expect some overlaps among the various geo-ontological domains—or so we will suggest.

---

[7] See Chaps. 4 and 5.

[8] See Chap. 3.

[9] In Tambassi and Magro (2015), Tambassi (2016), I named this kind of geo-ontologies "geomatics, topological and geometrical ontologies." Now, I think that "spatial ontologies" best capture the content of this category in the simplest way.

[10] The aim of this classification is to guide the reader through the main geo-ontologies of the contemporary debate, by analyzing their fundamental, common, and distinctive features, and by showing the overlaps between different geographical domains. Obviously, the list is not complete but includes the most discussed, reused, and relevant geo-ontologies, together with some non-strictly geographical ontologies in which some geographical aspects are described.

Fig. 6.1 Geo-ontological
tri-partition

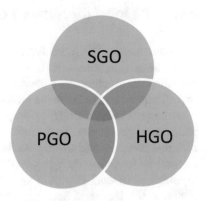

6.4.1   Spatial Geo-Ontologies

SGO are related to the computational processing of geographical data in GIS, GPS, and maps and are generally aimed at analyzing (spatially) Earth's surface, at locating (coordinates) and representing different geographic entities on maps, as well as at specifying the topological relations (disjunction, intersection, overlapping, inclusion, etc.) between these entities and the geometric aspects of the geographical investigation (elements like points, areas, solids, taxonomies, concepts, implicit and explicit geometries, and so on). A common feature of SGO is the high frequency of their (total or partial) repeated applicability in other ontologies. Such a repeated applicability is not surprising: the possibility of locating points, lines, and surfaces on a map is a recurring feature of geo-ontologies and a widespread need in many of their applications.

A prominent example of SGO is *WGS84 Geo Positioning*: a RDF vocabulary/ontology that provides the Semantic Web community with a namespace for representing geo-referenced positions and data in terms of altitude, longitude, and latitude. Another example is *GeoSPARQL*, which describes the geometric and topological aspect of geographic elements and their (mutual or absolute) position on the Earth's surface, by offering a taxonomy of geometries that range from the simple concept of point to complex geometries consisting of a plurality of solids. Geo-ontologies such as *Spatial Schema—ISO 19107, Schema for coverage geometry and functions—ISO 19123* and *Geography Markup Language (GML)—ISO 19136* provide a detailed inventory of concepts and properties direct to represent the geometrical aspects of geographical entities and their localization in a specific coordinate system. At the opposite side of these ontologies, as regards the theoretical richness of conceptualizations, we can find *Geometry (Ordnance Survey)*, which connects the concept of geometry (not further specified) with some elements that indicate the geometrical extents of some specific objects and areas. Finally, *NeoGeo Spatial Ontology* represents further topological relations between different geographic features.

6.4.2   Physical (or Natural) Geo-Ontologies

PGO are focused on those Earth aspects that are related to physical and natural phenomena (i.e., lithosphere, hydrosphere, atmosphere, pedosphere, biosphere, geomorphology, climatology, and so forth); they are numerically inferior to the SGO and strictly connected with the HGO.

Among PGO, *GEOSP—Geospecies* contains entities belonging to physical/natural geography. These entities represent a partition of the Earth's surface in different ecozones and aim to describe the geographical distribution of living species, as well as to define their habitats. *NDH Ontology (USGS)* and *Hydro Ontology (Spanish GeoData)* share the task of mapping the hydrological systems, respectively, of the USA and Spain, by producing a rich taxonomy of classes and properties, and by connecting those entities with some morphological elements. Given that hydrological systems are composed of both natural (lakes, rivers, etc.) and artificial (aqueducts, canals, etc.) entities, these ontologies also describe human artifacts and are therefore connected with HGO. Finally, *SWEET (Semantic Web for Earth and Environmental Terminology) ontologies* describe some aspects of Earth's geospheres, cryosphere, heliosphere, atmosphere, hydrosphere, land surface, ecological and physical phenomena, as well as their representations and transformations over time. *SWEET* ' classes and properties focus primarily on physical and natural features, point location (on maps), and climate change.

6.4.3   Human Geo-Ontologies

HGO deal with dynamics (e.g., historical and temporal modifications) and artifacts produced by political, administrative, social, urban, economical, population, cultural, archeological, historical, tourism, transportation geography, and so forth. The ontologies related to human geography constitute a numerically significant subgroup of the overall analyzed ontologies. This prominence is probably due to two different factors: the heterogeneity of the areas of research involved, and the fact that, being closely related to the human activity, HGO are strongly influenced by its organizing action. Their specificity, however, does not coincide with their repeated applicability, that is generally lower than that of other geo-ontologies.

FAO Geopolitical Ontology constitutes an excellent example of HGO: it has the scope of ensuring that the FAO and associated partners can rely on a master reference for geopolitical information, as it manages names in multiple languages, maps standard coding systems, provides relations among territories, and tracks historical changes. Such an ontology proposes a partition of (all) the geographical area types in territories and groups. The class of territories includes human entities such as nations and is divided into four subclasses based on the governmental autonomy of its members. An instance of that class is characterized in terms of location (coordinates), extension, GDP, human development index, population, and so forth. The class of groups has four subclasses too. Such subclasses represent organizations (FAO, UNESCO, etc.), geographic (Europe, North America, etc.) and economic (European Union, Arab Maghreb Union, etc.) areas, and special groups (i.e., SIDS: the organization of Small Island Developing States).

Other HGO such as *INSEE, Landinndelingen i Norge, The administrative geography and civil voting area ontology (Ordnance Survey)* and *Geopolitica (Spanish Geodata)* concern the political and administrative areas/subdivisions (and their historical and temporal modifications), respectively, of France, Norway, the UK, and Spain. *Vocabulario de Localizaciones* is used to model the physical locations of public places, *Postcode (Ordnance Survey)* aims to describe the postcode geography of Great Britain, *ISA Program Location Core Vocabulary* is specifically designed to aid the publication of data (names, addresses, locations) that is interoperable with EU INSPIRE Directive, *Transportes (Spanish GeoData)* describes transports domain and *NUTS* is a hierarchical system that divides up the economic territory of the EU, mainly for statistical and policy purposes.

6.4.4 Other Geo-Ontologies

The taxonomy that I have been proposing includes some overlaps among the three different geographical domains highlighted so far. Despite these overlaps, the specific aims of the aforementioned ontologies (see, in particular, *FAO Geopolitical Ontology, GEOSP—Geospecies, NDH Ontology* and *Hydro Ontology*) have allowed us to place them in a single exclusive category. The same cannot be said for ontologies such as *Erlangen CRM/OWL, Proton, LinkedGeoData, The Place Ontology, US Topographic*, and *GeoNames*. Some of which are parts of larger projects, where geography is (only) one of the different domains they analyze—for example, *Erlangen CRM/OWL* covers the entire area of cultural heritage and its list of geographical entities include especially entities related to geometry, topology, and location of places, but also implicitly connected to human and physical geography. Almost all of these ontologies range transversely across the three geographical domains just identified (SGO, PGO, and HGO) and propose geographical conceptualizations attempting to bring these domains together. In this sense, the multiplicity of geographical domains they analyze and their specific aims make the inclusion of these ontologies in a single and exclusive category arduous to be realized.

Figure 6.2 showcases one representative ontology for each kind of geo-ontology here suggested. In each case, it gives detailed information about objective, language, entities, and properties and overlaps of domains.

KIND OF GEO-ONTOLOGY	Spatial	Physical	Human	Other
NAME OF GEO-ONTOLOGY	WGS84 Geo Positioning	Hydro Ontology	FAO Geopolitical Ontology	Erlangen CRM/OWL
MAIN OBJECTIVES	Representing latitude, longitude and altitude information in the WGS84 geodetic reference datum	Describing hydrographical phenomena domain	Facilitating data exchange and sharing in a standardized manner among systems managing geopolitical information about countries and/or regions	Providing a formal structure for describing the implicit and explicit concepts and relationships used in cultural heritage documentation
LANGUAGE	RDF	OWL	OWL	OWL
(SELECTED) DISTINCTIVE ENTITIES	Points	Body of Water (subclass: Seawater, Fresh Water, Continental Water), Morphology (subclasses: Alluvional Soil, Cleft, Fluvial Island	Group (subclasses: Economic Region, Geographical Region, Organization, Special Group), Territories (subclasses: Disputed, Non Self Governing, Self Governing, Other)	Actor, Dissolution Event, Formation, Group, Place, Site, Place Appellation, Spatial Coordinates
(SELECTED) DISTINCTIVE PROPERTIES	Latitude, Longitude, Altitude	Concentration of Salinity, Coordinate, Is Tributary Of, Has Source From, Has Tributary, Flow Into,	Has Border With, Is Administered By, Is Predecessor Of, Is Successor Of, Has Coordinate, Has Nationality, Has Statistics	Dissolved, Has Formed, Is Located On Or Within, Occupies, Moved To, Moved From, Overlaps With, Borders With
OVERLAPS WITH OTHER GEOGRAPHICAL DOMAINS	None	Spatial Human	Spatial	Spatial Physical Human
IMPORTED GEO-ONTOLOGIES	None	WGS84 Geo Positioning FAO Geopolitical Ontology	None	None

Fig. 6.2 Schema of the main contemporary geo-ontologies

6.5 Conclusion

The aim of this chapter has been to provide a classification of geo-ontologies suitable for their spreading in the geographical debate.

As mentioned in Sect. 6.2, there are several taxonomies of ontologies in the literature of IT/computer sciences. However, none of them seems to be suitable for eliciting the specific geographical features of geo-ontologies. The same can be said for the taxonomy of *W3C Geospatial Incubator Group* that remains anchored to some specific aspects of geo-ontological analysis, rather than to geography as a discipline.

For this reason, my purpose has been to ground the classification of geo-ontologies on the fundamental geographical distinction between human and physical geography already embraced by the traditional debate. Moreover, since the most recurring (if not the only) feature of those ontologies is the possibility of locating points, lines, and surfaces on a map, we found it appropriate to introduce another type of geo-ontologies, the spatial ones, that conceptually traces what Pattinson defines as "spatial analysis" and Sala names "technical geography". The final tri-partition among spatial, physical, and human geo-ontologies represents an all-embracing classification that describes their main contents from a geographical point of view.

Obviously, this does not mean that there can be no geographical alternatives other than the taxonomy I have sketched. An idea, for example, could be to specify the kind of geography (classical or non-classical)[11] that lies behind geo-ontologies. Another one would be to decide whether their geographical conceptualizations denote a *lower* or a *professional*[12] geography.[13] This latter idea would allow for the categorization, among professional geo-ontologies, of what I place among SGO, as well as those geo-ontologies intended for a technical use (such as *NDH Ontology (USGS), Hydro Ontology (Spanish GeoData)* and some subclasses of *FAO Geopolitical Ontology*). Consequently, the other geo-ontologies can be included among the lower geo-ontologies.

Despite these alternatives, I think that the classification into SGO, PGO, and HGO might facilitate the introduction of geo-ontologies into the geographical debate for two reasons:

6.5.1. the first one is that the precise boundary between professional and lower geography (as well as between classical or non-classical geography) may be difficult to draw;

6.5.2. the second one is that, by capturing some fundamental geographical distinctions, such a classification might represent a useful tool to show how geo-ontologists conceptualize some specific geographical sub-areas, in terms of classes, individuals, and properties.

In addition, it can be observed that geo-ontologies might benefit from the interplay with the geographical and philosophical debate, too. Indeed, geography might

---

[11] See Chap. 3.

[12] See, for example, Egenhofer and Mark (1995), Geus and Thiering (2014).

[13] See Chap. 2 and Sect. 2.3.

gradually trace the guidelines for a classification within which the development of geo-ontologies will apply to all the geographical sub-disciplines by influencing the advancement of these ontologies in terms of professional/academic conceptualizations (see Chap. 1). As for the philosophical debate on geography, it might provide a specific framework aimed at discussing basic geographical notions and obtaining a more precise categorization of the entities conceived by contemporary geo-ontologies.

References

Battle R, Kolas D (2012) Enabling the geospatial semantic web with parliament and GeoSPARQL. Semantic Web 3(4):355–370

Berners-Lee T, Hendler J, Lassila O (2001) The semantic web. Sci Am 29–37

Bishr YA, Kuhn W (2000) Ontology-based modelling of geospatial information. In: Ostman A, Gould M, Sarjakoski T (eds) Proceedings of the 3rd AGILE conference on geographic information science, Helsinki, pp 24–27

Boll S et al (eds) (2008) LOCWEB '08: proceedings of the first international workshop on location and the web. ACM, New York

Buccella A, Perez L, Cechich A (2008) GeoMergeP: supporting an ontological approach to geographic information integration. In: International conference of the Chilean computer science society. http://disi.unitn.it/*p2p/RelatedWork/Matching/bucc-perbel-cech08p.pdf

Bullinger A (2008) Innovation and ontologies. Structuring the early stages of innovation management. Gabler, Wiesbaden

Câmara G, Monteiro A, Paiva J, Souza R (2000) Action-driven ontologies of the geographical space: beyond the field-object debate. In: GIScience 2000—program of the first international conference on geographic information science, Savannah, pp 52–54

Egenhofer M, Mark DM (1995) Naive geography. In: Frank AU, Kuhn W (eds) Spatial information theory: a theoretical basis for GIS. Proceedings of the second international conference. Springer, Berlin, Heidelberg, pp 1–15

Euzenat J, Shvaiko P (2007) Ontology matching. Springer, Heidelberg

Frank A (2001) Tiers of ontology and consistency constraints in geographical information systems. Int J Geogr Inf Sci 15(7):667–678

Geus K, Thiering M (2014) Common sense geography and mental modelling: setting the stage. In: Geus K, Thiering M (eds) Features of common sense geography. Implicit knowledge structures in ancient geographical texts. LIT, Wien

Gibson C (2009) Human geography. In: Kitchin R, Thrift N (eds) The international encyclopedia of human geography, pp 218–231

Gómez-Pérez A, Fernández-López M, Corcho O (2004) Ontological engineering: with examples from the areas of knowledge management, E-Commerce and the semantic web. Springer, Berlin, Heidelberg, New York

Goodchild M (2007) Citizens as sensors: the world of volunteered geography. http://www.ncgia.ucsb.edu/projects/vgi/docs/position/Goodchild_VGI2007.pdf

Guarino N (1998) Formal ontology and information systems. In: Proceedings of FOIS '98. Trento, Italy. IOS Press, Amsterdam, pp 3–15

Hudson-Smith A, Crooks A (2008). The renaissance of geographic information: neogeography, gaming and second life, CASA working paper 142, centre for advanced spatial analysis, London, University College London. http://www.bartlett.ucl.ac.uk/casa/pdf/paper142.pdf

Janowicz K (2006) Sim-dl: Towards a semantic similarity measurement theory for the description logic CNR in geographic information retrieval. OTM Workshops 2:1681–1692

Kavouras M, Kokla M, Tomai E (2005) Comparing categories among geographic ontologies. Comput Geosci 31(2):145–154

Kuhn W (2001) Ontologies in support of activities in geographical space. Int J Geogr Inf Sci 15(7):613–631

Kyzirakos K, Vlachopoulos I, Savva D, Manegold S, Koubarakis M (2014) GeoTriples: a tool for publishing geospatial data as RDF graphs using R2RML mappings. In; Proceedings of the 13th international semantic web conference, poster & demonstration track, pp 393–396

Lassila O, McGuinness DL (2001) The role of frame-based representation on the semantic web. Technical report knowledge systems laboratory, No. 01-02, Stanford

Mizoguchi R, Vanwelkenhuysen J, Ikeda M (1995) Task ontology for Reuse of problem solving knowledge. In: Mars NJ (ed) Towards very large knowledge bases—knowledge building and knowledge sharing. IOS Press, Amsterdam, pp 46–57

Pattinson WD (1963) The four traditions of geography. J Geogr 63(5):211–216

Perry M, Herring J (eds) (2012) OGC GeoSPARQL—a geographic query language for RDF data. OGC© Standard

Rodrìguez M, Egenhofer M (2004) Comparing geospatial entity classes: an asymmetric and context-dependent similarity measure. Int J Geogr Inf Sci 18(3):229–256

Sala M (2009) Geography. In: Sala M (ed) Geography. Encyclopedia of life support systems. EOLSS Publisher, Oxford, pp 1–56

Sowa JF (2000) Guided tour of ontology. http://www.jfsowa.com/ontology/guided.htm

Tambassi T (2016) Rethinking geo-ontologies from a philosophical point of view. J Res Didactics Geogr (J-READING) 2(5):51–62

Tambassi T, Magro D (2015) Ontologie informatiche della geografia. Una sistematizzazione del dibattito contemporaneo. Rivista Di Estetica 58:191–205

Turner AJ (2006) Introduction to NeoGeography. O'Reilly Media Inc., Sebastopol (CA). http://pcmlp.socleg.ox.ac.uk/sites/pcmlp.socleg.ox.ac.uk/files/Introduction_to_Neogeography.pdf

Uschold M, Grueninger M (1996) Ontologies: principles, methods and applications. Technical Report of the Artificial Intelligence Applications Institute, No. 191. Edinburgh, Scotland

van Heijst G, van der Spek R, Kruizinga E (1996) Organizing corporate memories. In: Gaines B, Musen MA (eds) Tenth knowledge acquisition for knowledge-based systems workshop (KAW 96). Banff, Canada, pp 42.1–42.17

Visser U (2004) Intelligent information integration for the semantic web. Lecture notes. In: Computer science, vol 3159. Springer, Heidelberg

Warf B, Arias S (eds) (2009) The spatial turn interdisciplinary perspectives. Routledge, London New York

Chapter 7
Ontological Categories for Geo-Ontologies

Abstract Despite their recent development (see Chap. 6), geo-ontologies still represent a complicated conundrum for most experts involved in their design. IT/computer scientists use ontologies for describing the meaning of data and their semantics to make information resources built for humans also understandable for artificial agents (see Chap. 2). Geographers pursue conceptualizations that describe the (geographical) domain of interest in a way that should be accessible, informative, and exhaustive for their final recipients (see Chaps. 2 and 6). In this context, philosophers/ontologists should offer conceptual solutions for carving nature at the joints and choosing how to categorize all the entities belonging to the geographical domain. This chapter aims to combine assumptions and requirements coming from these different areas of research, in order to show what categories might complete the current domain of geo-ontologies. The issue is approached by thinking of such a domain as a whole, composed of two different levels of categorization. The first level concerns the IT components shared among different ontologies. The second level deals with contents, for which (philosophical and) geographical analysis can include categories that do not appear at the first level.

Keywords Common-sense geography · Geographical conceptualization · Geo-ontologies · IT components · IT/computer ontologies · Levels of categorization · Ontological categories · Ontological contents

7.1 The Geo-Ontological Domain

Without claiming to be exhaustive, we can speak about geo-ontologies in terms of formal representations of geographical entities, aimed at describing geographical application domains or, at least, some of their very specific sub-areas. If such a definition might appear as vague, imprecise, or even tautological, we can turn to the different disciplinary areas which make up this specific ontological domain, namely IT/computer science, geography, and philosophy.[1]

---

[1] See Chaps. 1–2 and 6.

© The Author(s), under exclusive license to Springer Nature Switzerland AG 2021
T. Tambassi, *The Philosophy of Geo-Ontologies*, SpringerBriefs in Geography,
https://doi.org/10.1007/978-3-030-78145-3_7

In IT/computer science, where the spread of these formal representations can be ascribed to the development of the Semantic Web (Bishr and Kuhn 2000; Kuhn 2001; Buccella et al. 2008), an ontology can be thought of as an explicit (and sometimes partial) specification of a shared conceptualization that is formalized in a logical theory (Gruber 1993; Guarino and Giaretta 1995; Borst 1997; Studer et al. 1998; Tambassi and Magro 2015).

From a geographical perspective, ontology is the discipline that analyzes the mesoscopic world of geographical partitions by outlining what kinds of geographical entities exist, and how they can be classified and related in a system that gathers them together.[2]

Finally, in the philosophical (in particular, analytical) domain, the term "ontology" denotes the discipline concerned with the question of what entities exist. The task is often identified with that of drafting, at a high level of abstraction, a complete inventory of reality by specifying its hierarchical structure in terms of categories (of being) and relations among them. To be more precise, it is claimed that for covering all the entities in the world, we should refer to some (ontological) categories which have to include everything that exists.[3] But, what do we mean when we speak about "(ontological) categories"? And, more in general, what are the issues emerging from a categorial ontology?

7.2   Issues of a Categorial Ontology

Getting a proper meaning of "ontological category" may be complicated. Wester-hoff (2005), for example, distinguishes two main theories, depending on whether one is willing to bypass the problem or to take it at face value. According to the former, when asked to provide a definition of ontological category, we can alternatively declare it an impossible enterprise (Grossmann 1983), give only examples of categories and highlight our immediate acquaintance (Hartmann 1949), or provide an entire laundry list of (all of) them without further specifications. Apart from these theories, which evade rather than answer the question, Westerhoff also distinguishes three kinds of accounts which try to deal with the definition of ontological category in a systematic way. The first one maintains that ontological categories are the most general kinds of things. Accordingly, an ontological category is more general than other categories, and the relationship among ontological categories and other categories can be expressed, alternatively, in terms of sets (Rosenkrantz and Hoffmann 1991) or ontological dependence (Norton 1976). The intuition behind the second account is that entities belonging to the same ontological category can somehow be exchanged (or intersubstituted) for one another in certain contexts. Such an intuition explains why certain substitutions in a statement make the statement false, while

---

[2]See Chap. 2. For an analysis on the "multifaceted nature" of the relationship between ontology and geography, see Chap. 1, Sect. 1.3.

[3]See Chap. 2.

others make it meaningless (Ryle 1949; Sommers 1963, 1971). Finally, the third account maintains that ontological categories provide the identity criteria for classes of entities. Thus, two entities belong to the same category if their criteria of identity are the same (Frege 1884; Dummett 1981).

Surely, the question of the definition does not exhaust the issues emerging from a categorial ontology that might be divided into two different sorts: form and content. Formal (or structural) issues generally involve the hierarchical organization of our ontology and the relationships among different categories. Otherwise, issues concerning contents regard what sorts of ontological categories should be considered as fundamental and what entities should be included among them. The most prominent examples of (fundamental) categories embrace entities such as objects, individuals, properties, particulars, attributes, relations, states of affairs, modes, tropes, facts, events, processes, and so forth. Usually, the basic categories are identified on empirical and cognitive grounds and represent a theoretical compromise between two different aims: cognitive economy and explanatory power (for which the notion of causality plays a fundamental role).

7.3   On the General Aims of Geo-Ontologies

In order to show what different categories might complete the domain of geo-ontologies, it may be useful to start by sketching the main aims of geo-ontologies that can be separated into two different sorts. On the one hand, there are the specific aims that reflect the particular purpose for which a geo-ontology has been created—as well as the point of view of the community sharing the ontology in question. On the other hand, general aims (mostly shared by the stakeholders of the contemporary debate) are essentially four:

7.3.1.   repeated applicability;
7.3.2.   informativeness;
7.3.3.   completeness;
7.3.4.   accessibility.

"Repeated applicability" refers to the process in which existing ontological knowledge is used as input to generate new ontologies in the Semantic Web. Such a process increases the quality of the applications using ontologies, as these applications become interoperable and are provided with a deeper, machine-processable and commonly agreed upon understanding of the underlying domain of interest. Moreover, repeated applicability allows the integration and aggregation of data and information and avoids the re-implementation of ontological components that are already available on the Web (Pâslaru-Bontaş 2007).

"Informativeness" indicates the need to disclose and organize the ontological contents in a meaningful way for the final recipients (specialists and/or non-expert users) of the ontology. This contributes to enhance a mutual understanding between different communities of human beings, as well as the communication between human beings and software systems and between software systems themselves (Goy and Magro 2015).

"Completeness" refers to the quantity of information belonging to the ontologies and points out that such an information should be detailed and exhaustive for the domain that we want to represent—all this despite the fact that the domain in question can be (restricted or) partial and might be represented with a specific perspective or a particular viewpoint.

7.4   From the Need of Accessibility to the Common-Sense Geography

The last general aim of geo-ontologies is represented by "accessibility," which indicates that the information they express should be understandable and usable for a variety of users—including the scientific community and, in particular, non-experts. A useful tool in this respect is the common-sense geography (CSG) that can be defined as the body of knowledge, theories, and beliefs that people have about the surrounding geographical world (Egenhofer and Mark 1995). To be more precise, CSG denotes a *lower* (or small-g)[4] geography (to be distinguished from the *higher* or academic geography): that is, "the phenomenon of the spread and application of geographic knowledge outside of expert circles and disciplinary contexts" (Geus and Thiering 2014).

The idea behind CSG is to establish a connection between how people think about geographic space and how to develop formal models of such reasoning that can be incorporated and integrated into software systems (Egenhofer and Mark 1995). According to Smith and Mark (2001), the development of that connection allows the transformation of quantitative geospatial data into the sorts of qualitative representations of geospatial phenomena that are tractable to non-expert users. Moreover, CSG might help us also in our efforts to maximize the usability of corresponding information systems, rendering the results of work in geospatial ontology compatible with the results of ontological investigations of neighboring domains.

CSG is generally organized in systems of entities falling under categories (of kinds, types or universals), typically determined by prototypical instances. Hierarchically, these systems take the form of a tree: they have one all-embracing category at the top level, with more general categories at the subsequent levels and more specific categories as we move down each of the various branches. As for the contents, GSG's primary axis is a system of objects, while attributes (properties, aspects, features), relations, events, and so forth form a secondary axis of the ontology. The primacy of objects depends on the fact that, in this context, attributes, relations, and events are, respectively, attributes of, relations between and events involving objects, in ways

---

[4]See Chap. 1.

that imply a dependence of entities in these latter categories upon their hosts in the primary category of objects (Smith and Mark 2001).[5]

7.5   Ontology Components

With all these conceptual assumptions, let us go further with ontological contents, by sparing a few words for the core components shared among different IT/computer ontologies. According to Lord (2010) and Laurini (2017), such components are essentially three:

7.5.1.   classes;
7.5.2.   instances;
7.5.3.   relations.

"Classes" (also known as concepts, kinds, frames, or types) represent groups or sets of different instances that share common features. They can be defined by extension (enumerating their instances) or by intension (giving restrictions to their instances) (Jaziri and Gargouri 2010). Classes can also contain (more specific) subclasses and/or be subclasses of other (less specific) classes. This means that if the class A is a subclass of B, then any instance of A is also an instance of B. Moreover, classes can share relations with each other: such relations generally describe the way in which instances of one class relate to the instances of another.

"Instances" (also called individuals or particulars) are the lowest-level components (the base units) of an ontology and may model concrete objects such as rivers or deserts, or more abstract objects such as countries, regions, or functions (Laurini 2017).

Finally, "relations" describe the way in which classes and instances relate to and interact with each other. To be more precise, relations can normally be expressed directly between instances or between classes of the domain and may be distinguished according to the number of classes related: reflexive relations link only one class, binary relations link two classes, and n-ary relations link more than two classes.

Otherwise, Jaziri and Gargouri (2010) also include among the main components of IT/computer ontology "slots," which describe the various features of a class and its instances (Noy and McGuinness 2003). More precisely, slots (also known as properties, attributes, or roles) "contribute to identify [classes] by characterizing them and can be used in intensional definitions of [classes], to relate [instances] or to give attribute values" (Jaziri and Gargouri 2010, p. 38). Finally, slots allow for the expression of relationships among classes in a domain, such as hierarchy and consequently constitute the basis for the hierarchical structure of the ontology.

---

[5] See Chap. 2.

7.6   Between Formal Ontologies and Ontological Categories

Noy and McGuinness (2003) maintain that developing an IT/computer ontology requires (at least) four different steps:

7.6.1.   defining classes in the ontology and individual instances of these classes;
7.6.2.   arranging the classes in a taxonomic hierarchy;
7.6.3.   defining slots and describing allowed values for these slots;
7.6.4.   filling in the values for slots for instances.

Now, if we combine such steps with the theoretical considerations expressed in the previous sections, we can briefly summarize what we need for developing a geo-ontology as follows.

From a formal perspective, a hierarchical tree structure that classifies classes of different instances is required. Such a structure should fulfill the needs expressed by 7.6.1 and 7.6.2 with the aim of accessibility specified by the considerations on CSG.

As for contents, we can easily observe that a geo-ontology should include categories such as classes (kinds) and instances (objects). But the inclusion can also be extended to entities like relations and slots (properties), which we have counted among the core components of IT/computer ontologies. More precisely, relations and slots shall, respectively, link and characterize both classes (kinds) and instances (objects), as we have remarked in Sect. 7.5. However, relations and slots cannot constitute the primary axis of our geo-ontology since, according to CSG, only the category of instances can fill that role. Remarking on their ontological dependence from the category of instances does not mean denying their fundamentality in a geo-ontology, nor their specific role for describing *exhaustively* the geographical domain of interest, as the aim of completeness requires.

Finally, we should add that the contents of a geo-ontology must be informative both for specialists and non-expert users. Accordingly, as Cumpa (2014) suggests, the ontological categories shall have the explanatory power to account for the geo-ontological domain as a complex composed of a scientific level with which specialists/geographers are acquainted in their research *and*, in particular, an ordinary level of thinghood emerging from commonsensical geographical experience.

7.7   Lowe's Four-Category Ontology

If the above considerations seem to provide some clear guidelines for the (hierarchical) structure and contents (in terms of ontological categories) of geo-ontologies, it is interesting that, quite independently of the geo-ontological debate, the definitive version of Lowe's ontological proposal (2006) exhibits significant similarities.

According to the author, his four-category ontology provides an *exhaustive* inventory of what there is and an explanatory framework for a metaphysical foundation for

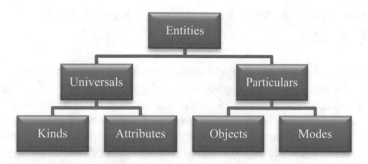

Fig. 7.1 Lowe's four-category ontology

natural science, taking the common sense as a starting point.[6] By "the four-category ontology," he means a system that identifies four fundamental ontological categories (objects, kinds, attributes, and modes) that result from the exhaustive and exclusive distinctions between universals and particulars and between substantial and non-substantial. The fundamentality of these categories is expressed by the fact that the existence and identity conditions of the entities belonging to such categories cannot be exhaustively specified in terms of ontological dependency relations between those entities and the entities belonging to other categories (Lowe 2006, p. 8).

Lowe's fundamental categories are related to one another in a hierarchical tree structure (see Fig. 7.1), with the topmost level the category of entities to which anything whatsoever belongs. The concept of instantiation captures the division between universals and particulars at the second level: particulars instantiate universals, while universals are instantiable. At the third level, the distinction between substantial and non-substantial allows us to divide universals into the two sub-categories of substantial universals (*kinds*) and non-substantial universals (*attributes*), and particulars into the sub-categories of substantial particulars (*objects*) and non-substantial particulars (*modes*). Kinds are those universals that have objects as their instances. Attributes are those universals that have modes as their instances. Objects are characterized by modes. Kinds are characterized by properties and relations. From the point of view of contents, attributes include properties and relations, conceived as universals. Kinds include as paradigm examples natural kinds of persisting objects. Modes include property and relation instances, otherwise known as monadic and relational tropes. Substantial particulars include propertied individuals, the paradigm examples of which are persisting, concrete objects. Such a category occupies a more fundamental place in the scheme of being. This is due to the fact that, according to Lowe, objects are the only independent entities and for this they are ontologically prior to other categories.

---

[6]However, Lowe himself also acknowledges that there are aspects of common sense that need to be revised or abandoned.

7.8 Overlaps and Deviations: Cumpa's Fact-Oriented Ontology

Despite the similarities, the overlap between geo-ontologies and Lowe's proposal should not be considered as complete and totally consistent. First, Lowe includes (universal) relations and properties within the ontological category of attributes, whereas geo-ontologists divide these two kinds of entities into the core components of relations and slots. Second, the category of modes occupies a fundamental place in Lowe's ontology, while in geo-ontologies such a category can be thought of as a subclass of relations. Third, if common sense might be thought of as a requirement for geo-ontologies' need of accessibility, it only represents a starting point—that can be revisited or abandoned—in Lowe's proposal. For these reasons, the previous considerations should be interpreted as an endorsement of neither Lowe's four-category ontology nor other ontological systems that capture our commonsense conceptual schemas by including two fundamental categories of particulars (substances and accidents) and two universals (kinds and property universals).[7]

Moreover, such considerations cannot even be regarded as endorsing Cumpa's overall account (see Sect. 7.6). For his proposal aims at meeting the needs of accessibility and informativeness of geo-ontologies (by considering the world as a complex of, respectively, ordinary and scientific levels of thinghood that categories should have the explanatory power to describe), at the same time his "fact-oriented ontology" seems to be incompatible both with the components of (geo-)ontologies (see Sect. 7.5) and with the system of objects that CSG assumes. But then, what kind of ontological categories best represent the domain of geo-ontologies? In approaching such a question, an idea would be to maintain the basics of Cumpa's dichotomy, modifying its horns: that is, to think about the overall geo-ontological domain composed of two different levels of categorization, one which concerns *(core) components* and another which deals with *contents*.

7.9 From Components to Contents

The first level of categorization is devoted to the analysis of core components, by describing how shared conceptualizations (whatever conceptualization) are represented *in an IT/computer system*. Therefore, such a level is closely related with the domain of IT/computer science. And given that IT/computer scientists currently identify four different core components of ontologies (classes, instances, relations, and slots), the ontological categorization at this level should follow such a guideline, respecting the limits set by the advancements of IT/computer science. This does not mean that the core components (as well as general aims of ontologies) will never change, and so will the ontological categories used to describe all these changes.

---

[7] See for example Smith (1997).

The second level concerns ontological contents, in particular, it aims at providing (different) representations of the geographical world, where the plural of the term "representation" reflects both the particular purpose for which a geo-ontology has been created and the specific geographical domain that has to be described. At this level, although IT/computer science constitutes the main background of the geo ontological categorization, philosophical and geographical analysis has more *theoretical freedom* to elaborate different categorizations of the geographical world. That means that the categorization of contents does not need to correspond with the ontological categories at the first level. Therefore, such a categorization might include any kind of ontological category (e.g., modes, facts, events, and so forth) based on the (general as well as specific) aims of a geo-ontology and on the domain we want to represent.

The two levels are not, however, totally independent: In particular, the second level might be conceived as dependent upon the first one. Indeed, we could hardly think about categorizing the entities in this specific context, without an IT/computer structure that tells us *where* to categorize entities. And *where*, here, indicates the components specified at the first level. *Where* to categorize an entity does not correspond to *how* to categorize such an entity. This means that while—at present—the ultimate structure, along with the fundamental categories of our geo-ontology, should be conceived as fixed, the same cannot be said both for the location of the entities in such a structure and for the list of entities that we include in our geo-ontology. Surely, the aim of accessibility of geo-ontologies is best achieved by a categorial correspondence of the two levels. However, even such a correspondence would not imply the ultimate exclusion, for example, of the category of facts, which could rather be important for improving the informativeness (and maybe the completeness) of our geo-ontology. Nevertheless, as such a category is not included among the core component of geo-ontologies, it cannot be considered as fundamental in this context, and, therefore, its entities should be ultimately located among one (or more) of the core components of (geo-)ontologies.

7.10 Variantism

In commenting on this categorial proposal, Cumpa emphasizes that "a fundamental fact about geo-ontologies is that structures of categories can change due to certain practical preferences" (Cumpa 2019, p. 149). Accordingly, the author labels my position on geo-ontologies "structure variantism," meaning that the structures of ontological categories (the second level of categorization) can change for certain practical preferences. Then, Cumpa draws attention to what he maintains is a very specific issue of my position, by addressing the following question: how can the four fundamental categories at the first level (instances, classes, relations, and slots) be invariant and the structures formed by them be variant?

In answering such a question, I have to spare a few more words on the first level of categorization. In my proposal, this level is closely related to the domain of IT/computer science, which provides the IT structures of geo-ontologies, namely their core components. I argue that, at present, the debate in IT/computer science generally identifies four core components of geo-ontologies: instances, classes, relations, and slots. However, this does not exclude that advancements of IT/computer science could expand, reduce, or modify the list of core components. In this sense, it is only in respect to the contemporary debate that the core components of geo-ontologies should be conceived as invariant. Consequently, I totally agree with Cumpa, who considers my position a form of "structure variantism". However, I would also add that such a variantism is not just about the structure, it is also about the fundamental categories. It is a "categorial and structural variantism" and implies that both fundamental categories and structure can change according to, on the one hand, the advancements of IT/computer science, and on the other hand, the purpose for which a geo-ontology has been created and the geographical domain that has to be described (see Sect. 7.9). In this sense, there is no contradiction between invariantism of categories and variantism of structure, since in my view both categories and structure are variants.

7.11   Two Kinds of Completeness

But Cumpa's criticism of my proposal is not limited to the fundamental categories of geo-ontologies. It also involves their aim of completeness, which I sketched in Sect. 7.3.

About this aim, Cumpa underlines that an uncritical belief silently encompassing the categorial debate (from Plato to contemporary substantialists) is that we should, in some way, justify the completeness of any system of ontological categories. The ground of this belief, he states,

> is the difficulty Aristotelian substantialism faces in accounting for relations among (enumerated) categories. This seems to mean that the inquiry into the categorial completeness of systems of categories is not a problem for categorial ontology in general, but rather just a problem for the Aristotelian tradition. (Cumpa 2019, p. 152)

On the contrary, since factualism (that he supports) considers the world not as an enumeration but as a combination of categories, it need not deal with this categorial problem.

Then, Cumpa's reflection on categorial completeness questions my position on facts, by focusing on two main points:

7.11.1.   the first one is that I am sympathetic to the idea that facts might help improve the aim of completeness (and informativeness) of geo-ontologies;

7.11.2.   the second one is that, although I do not exclude the category of facts from the categories of geo-ontology, I also argue against the fundamentality of such a category.

Finally, the issue that Cumpa raises is: If categorial completeness is one of the essential aims of geo-ontologies and facts help to account for it, how can facts still be a non-essential category?

Before answering such a question, I wish to point out my uncertainty about the fact that the notion of completeness "silently encompassing the categorial debate" corresponds to the aim of completeness of geo-ontologies. (I do not even think that Cumpa supports such a correspondence.) In fact, while the first kind of completeness refers to the (structures formed by) relations among (and the total number of) ontological categories, and the second one concerns specifically the (geo-)ontological debate. More precisely, I talk about completeness in geo-ontologies, by underlining that information belonging to a geo-ontology should *exhaustively* describe its domain of interest. This means that if we want to develop a geo-ontology that represents the bodies of water in Italy, the geo-ontological aim of completeness is fulfilled if and only if no (Italian) body of water is excluded from the ontology.[8] Broadly speaking, we could claim that the first kind of completeness (Cumpa's) concerns the *types* of categories, while the second one primarily refers to their *tokens*.

Assuming such a distinction concerning the two notions of completeness does not mean:

7.11.3.   the denial of any connection between them;
7.11.4.   the reduction of the potential soundness of Cumpa's criticism to the Aristotelian tradition (Lowe 2006; Heil 2012);
7.11.5.   the rejection of the fact that the aim of completeness of geo-ontologies is somehow related to the first kind of completeness;
7.11.6.   the lessening of the validity of Cumpa's criticism to my reflection on completeness.

Here, I do not intend to take a position regarding the first three points of the previous list, but rather to answer 7.11.6. I thus try to respond to the following questions: Do facts really improve the aim of completeness of (geo-)ontologies? If yes, should we include the category of facts within the categories of (geo-)ontologies? If this is the case, should we also consider the category of facts as fundamental? Might the category of facts be non-fundamental even if it improves the aim of completeness of (geo-)ontologies?

7.12 Completeness and Fundamentality

In my view, if we consider "the *great fire of Rome* of *64* AD" as a fact and we have to build a historical ontology of ancient Rome, we cannot exclude such a fact from the ontology, especially if we want to describe *exhaustively* the domain of this historical ontology. In other words, if an entity improves the completeness (of the contents)

---

[8]For a deep investigation on the notion of completeness in IT/computer ontology, see Bittner and Smith (2008).

of a certain ontology, then such an entity should be included within that ontology. But if the completeness of a geo-ontology benefits from the inclusion of such a fact, why could we not also include the category of facts among the categories of a geo-ontology? I have no objection to that. Indeed, at the second level of categorization we can, in principle, insert any kind of category, including facts. However, such an inclusion does not mean considering the category of facts as fundamental. Why? Because, IT/computer science, at present, just identifies four core components of geo-ontologies (instances, classes, relations, and slots), which represent their fundamental categories. Facts are not on the list. Nevertheless, such an exclusion does not prevent advancements in computer science from allowing the *future* inclusion of the category of facts among the core components of geo-ontologies. This means that, according to my proposal, the possibility of including facts at the first level of categorization is jurisdiction of IT/computer scientists (and IT advancements), not of philosophers.

But then, if facts are not core components of geo-ontologies, how do we include a fact such as "the *great fire of Rome* of *64* AD" within an ontology? In my view, we could, for example, consider this fact as a property (*slot*) of Rome in a specific period of time, as an *instance* of the class "fires", as a *relation* among the entities "Rome", "Fires", "64 AD", and so forth. Indeed, in Sect. 7.9, I maintain that while, at present, the fundamental categories (the core components) of geo-ontologies should be conceived as fixed, the same cannot be said either for the entities we want to include *within* the core components or for the location of the entities in question.

Now, if the exclusion of the category of facts among the core components may seem a trouble for the geo-ontological aim of informativeness[9]—we could for instance ask why we should consider a fact as belonging to another category—we can simply add that:

7.12.1.   a balance between informativeness and completeness is never easy;
7.12.2.   the informativeness of an ontology might sometimes be at the expense of the completeness (and maybe vice versa).

Finally, I also want to remark that there is nothing to stop us from filling up, for example, the *whole* core component of instances[10] with entities such as "the *great fire of Rome* of *64* AD"—in other words, with entities we are willing to consider as facts. In my view, there is no contradiction in this filling. First, nothing impedes us from also including other kinds of entities within such a category, so that facts are not the only examples of instances. Second, we must not forget that any entity belonging to a geo-ontology is ultimately categorized among one of the core components, no matter what kind of entity it is.

---

[9] See Sect. 7.3.

[10] The same reasoning can be easily extended to the other core components.

7.13 Philosophical Ontology Versus IT/Computer Ontology

In Chap. 2 and in Sect. 7.6, we have seen that Cumpa questions the relationship between ordinary world and physical universe. By "ordinary word", he understands "an ordinary level of thinghood with which ordinary people are acquainted in their commonsensical and practical experiences." By "physical universe", he means "a scientific level of thinghood with which scientists are acquainted in their experimental research, such as fundamental physics, chemistry, or biology" (Cumpa 2014, pp. 319–20). According to the author, the ordinary world and the physical universe are levels of thinghood, which are not isolated from each other. Thus, they require some ontological categories that have the explanatory power to account for the world as a complex composed of both ordinary entities and entities of fundamental sciences. The two horns of this relationship represent, I think, the (main) framework within which Cumpa develops his own ontological proposal.

Within this framework, the *theoretical freedom* of choosing what kinds of ontological categories best describes the world as a complex is bounded by people's experience *and* scientists' experimental research on the world itself. The framework of geo-ontologies is quite different, since it comprehends an area of investigation that ranges from IT/computer science to philosophy and geography. In this context, the categorial freedom is *also* bounded by the advancements of IT/computer science, which currently systematizes the entity of (geo-) ontologies into four main core components: instances, classes, relations, and slots. Within this categorial tie, the theoretical freedom generally concerns how best to categorize the entity among the core components of a geo-ontology, according to the specific purpose for which a specific geo-ontology has been created, its domain of investigation, the point of view adopted during its development, and the general aims of completeness, informativeness, accessibility, and repeated applicability.

The fact that our frameworks are different does not suffice—or so I believe—to (entirely) explain the differences between Cumpa's proposal and mine, which might be, respectively, labeled as sympathetic with reductionism and perspectivism (see Chap. 5). We should also bear in mind that, while Cumpa's proposal falls within philosophical (big-O) Ontology, my position should also be seen in connection to the debate on IT/computer (small-o) ontology. Obviously, this does not mean that different positions in different debates cannot be relevant to one another, *bringing up problems* and inviting answers to unexpected questions. But this also does not mean that, if the tables were turned, Cumpa would have maintained my thesis on geo-ontologies, nor that I would have *totally* embraced his factualist position on the fundamental categories (Cumpa 2014, 2018).

References

Bishr YA, Kuhn W (2000) Ontology-based modelling of geospatial information. In: Ostman A, Gould M, Sarjakoski T (eds) Proceedings of the 3rd AGILE conference on geographic information science, Helsinki, pp 24–27

Bittner T, Smith B (2008) A theory of granular partitions. In: Munn K, Smith B (eds) Applied ontology. An introduction. Ontos-Verlag, Berlin

Borst WN (1997) Construction of engineering ontologies, centre for telematica and information technology. University of Twente, Enschede

Buccella A, Perez L, Cechich A (2008) GeoMergeP: supporting an ontological approach to geographic information integration. In International Conference of the Chilean computer science society. http://disi.unitn.it/*p2p/RelatedWork/Matching/bucc-perbel-cech08p.pdf

Cumpa J (2014) A materialist criterion of fundamentality. Am Philos Q 51(4):319–324

Cumpa J (2018) Factualism and the scientific image. Int J Philos Stud 26(5):669–678

Cumpa J (2019) Structure and completeness: a defense of factualism in categorial ontology. Acta Analytica 34(2):145–153

Dummett M (1981) Frege: philosophy of language. Duckworth, London

Egenhofer M, Mark DM (1995) Naive geography. In: Frank AU, Kuhn E (eds) Spatial information theory: a theoretical basis for GIS. Proceedings of the second international conference. Springer, Berlin, Heidelberg, pp 1–15

Frege G (1884) Die Grundlagen der Arithmetik: Eine logisch-mathematische Untersuchung über den Begriff der Zahl. Koebner, Breslau

Geus K, Thiering M (2014) Common sense geography and mental modelling: Setting the stage. In: Geus K, Thiering M (eds) Features of common sense geography. Implicit knowledge structures in ancient geographical texts. LIT, Wien

Goy A, Magro D (2015) What are ontologies useful for? Encyclopedia of information science and technology. IGI Global, pp 7456–7464

Grossmann R (1983) The categorial structure of the world. Indiana University Press, Bloomington

Gruber TR (1993) A translation approach to portable ontology specifications. Knowl Acquis 5(2):199–220

Guarino N, Giaretta P (1995) Ontologies and knowledge bases—towards a terminological clarification. In: Mars NJ (ed) Towards very large knowledge bases—knowledge building and knowledge sharing. IOS Press, Amsterdam, pp 25–32

Hartmann N (1949) Der Aufbau der realen Welt: Grundriß der allgemeinen Kategorienlehre. Anton Hain, Meisenheim

Heil J (2012) The world as we find it. Oxford University Press, Oxford

Jaziri W, Gargouri F (2010) Ontology theory, management and design: an overview and future directions. In: Gargouri F, Jaziri W (eds) Ontology theory, management and design: advanced tools and models. Information Science Reference, Hershey, PA

Kuhn W (2001) Ontologies in support of activities in geographical space. Int J Geogr Inf Sci 15(7):613–631

Laurini R (2017) Geographic knowledge infrastructure: applications to territorial intelligence and smart cities. ISTE-Elsevier, London

Lord P (2010) Components of an Ontology. http://ontogenesis.knowledgeblog.org/514

Lowe EJ (2006) The four-category ontology: a metaphysical foundation for natural science. Clarendon Press, Oxford

Norton BG (1976) On defining 'ontology.' Metaphilosophy 7:102–115

Noy NF, McGuinness DL (2003) Ontology development 101: a guide to creating your first ontology. Stanford University, Stanford

Pâslaru-Bontaş E (2007). A contextual approach to ontology reuse. Methodology, methods and tools for the semantic web. Ph.D. Thesis, Department of Mathematics and Computer Science, Freien Universitat, Berlin

Runggaldier E, Kanzian C (1998) Grundprobleme der analytischen Ontologie. Verlag, Paderborn

Ryle G (1949) The concept of mind. Hutchinson's University Library, London

Smith B (1997) On substance, accidents and universals: in defence of a constituent ontology. Philos Pap 27:105–127

Smith B, Mark DM (2001) Geographical categories: an ontological investigation. Int J Geogr Inf Sci 15(7):591–612

Sommers F (1963) Types and ontology. Philos Rev 72:327–363

Sommers F (1971) Structural ontology. Philosophia 1:21–42

Studer R, Benjamins VR, Fensel D (1998) Knowledge engineering: principles and methods. IEEE Trans Data Knowl Eng 25(1–2):161–197

Tambassi T (2018) The riddle of reality. In: Tambassi T (ed) Studies in the ontology of E.J. Lowe. Editiones Scholasticae, Verlag

Tambassi T, Magro D (2015) Ontologie informatiche della geografia. Una sistematizzazione del dibattito contemporaneo. Rivista Di Estetica 58:191–205

Westerhoff J (2005) Ontological categories. Clarendon Press, Oxford

Conclusion

In the *Introduction*, it was said that this book pursues three main goals. The first one is to provide an overview of the ontological background and of the mutual interactions among the different areas of research that applied ontology of geography involves, namely IT/computer ontology, philosophical ontology, and geography. The second goal is to make explicit the ontological assumptions and commitments of geography in terms of spatial representations, geographical entities, and boundaries. The third one is to propose a geographical classification of geo-ontologies and to discuss what ontological categories might systematize their contents.

The three parts of this book, *Among Computer Science, Philosophy, and Geography: An Ontological Investigation, Systematizing the Geographical World* and *The Philosophy of Geo-ontologies*, pursue such goals and, more in general, explore the domain of applied ontology of geography: that is, the discipline concerned with the theoretical and technical needs of geographical/IT tools such as GIS and geo-ontologies, and with the ontological investigation behind geography as a discipline. While by analyzing the ontological background of applied ontology of geography, the first part of the book has been primarily expository, the other two have shown some theoretical solutions to account for the complexity of the geographical reality, as well as for its representation into geo-ontologies. Such theoretical solutions have been, respectively, named "perspectivism" and "variantism".

Perspectivism, in this context, points out that a geo-ontological investigation based on notions such as spatial representations, geographical entities, and boundaries cannot disregard the following factors. (1) There exist multiple conceptualizations of the geographical world. (2) Different languages and cultures slice such a world in different ways. (3) The geographical world has changed and will change over time. (4) Geography (as a discipline) has changed and will change over time too, by modifying its perspective, tools, domains of investigation, and aims. Therefore, what was, is, and will be considered as non-geographic could be considered as geographic and vice versa. (5) There were, are, and will be different kinds of geographies as well as different geographical branches, each of which had, has, and will have different tools,

© The Author(s), under exclusive license to Springer Nature Switzerland AG 2021
T. Tambassi, *The Philosophy of Geo-Ontologies*, SpringerBriefs in Geography,
https://doi.org/10.1007/978-3-030-78145-3

aims, and vocabularies. (6) The introduction of new scholarly fields and new technologies, the birth of intellectual movements, or paradigm shifts can/will influence geography as a discipline.

Variantism is stemming from the fact that, to show what categories might complete the domain of geo-ontologies, we should think of the domain as composed of two different levels of categorization. The first level is closely related to the domain of IT/computer science, which currently provides four core components of geo-ontologies: instances, classes, relations, and slots. The second level deals with contents, for which philosophical and geographical analysis might include any kind of ontological category. In this context, variantism indicates that core components and contents of geo-ontologies should not be conceived as fixed; quite the opposite, their categorizations can vary according to, on one side, the advancements of IT/computer science and, on the other side, the purpose for which a geo-ontology has been created and the geographical domain to describe.

Although perspectivism and variantism refer to different aspects of the geo-ontological investigation, these theoretical positions should not, however, be considered as isolated from each other, but as mutually related. Indeed, both perspectivism and variantism start from the assumption that there are multiple, alternative, and overlapping views on geographical reality, which can be represented and sliced in different ways. Moreover, both share the idea that applied ontology of geography should provide some platforms for describing, comparing, and maybe integrating the alternative views on geographical reality. In this sense, the task of applied ontology of geography is practical in nature and is subject to the same practical constraints experienced in all scientific activity. Consequently, any geo-ontological investigation should always be considered as a partial and imperfect edifice subject to correction and enhancement, open to new scientific needs and changes in the geographical world as well as in our ways to systematize it.

Index

A

Accessibility, 95, 110, 112, 114, 115, 119
Action, 30, 31, 56, 63, 69, 86, 100
Advancements, 21, 82, 86, 104, 114, 116, 118, 119, 124
Applied ontology of geography, 3, 6, 8, 21, 25, 63, 97, 98, 123, 124
Assumptions, 7, 12, 15, 25, 27, 29, 41, 45, 49, 58, 86, 87, 97, 107, 111, 123, 124
Attributes, 24, 31, 109–111, 113, 114
Axioms, 25, 39, 41–44, 96

B

Beliefs, 29–31, 49, 51, 53, 55–60, 72–74, 82, 110
Boundaries
   bona fide, 51, 54, 60, 71
   cultural, 49, 51, 56–58, 60
   fiat, 51–54, 56, 58–60, 71
   geographical, 49–51, 53–60, 71, 72, 80
   institutional, 54, 71
   mathematical, 52
   physical, 50, 54, 60, 71
   spatial, 51, 52

C

Cartography, 68, 84
Categories, 10, 24, 26–29, 31–34, 66, 71, 74, 75, 78, 79, 85, 96, 107–110, 112–119, 123, 124
Categorization, 8, 49, 51–53, 56, 58–60, 71–75, 79, 82, 96, 103, 104, 107, 114, 115, 118, 124
Causality, 31, 109

Classes, 24, 25, 34, 74, 81, 84, 95, 100, 103, 109, 111, 112, 114–116, 118, 119, 124
Classification, 7, 25, 49–51, 53–55, 57–60, 71–73, 75, 76, 93, 95–98, 103, 104, 123
Common sense, 21, 27–30, 39, 110, 113, 114
Communication, 9, 22, 24, 50, 94, 95, 109
Completeness, 85, 95, 109, 110, 112, 115–119
Components, 15, 69, 83, 107, 109, 111, 112, 114–116, 118, 119, 124
Conceptualizations, 9–12, 15, 16, 21–25, 29–31, 34, 63, 69, 85–87, 93, 95–97, 99, 101, 103, 104, 107, 108, 114, 123
Contents, 5, 22, 55, 87, 95, 97, 98, 103, 107, 109–115, 117, 123, 124
Cultures, 30, 56, 58, 63, 65, 73, 80, 83, 85–87, 94, 123

D

Data, 9, 21, 22, 28, 34, 56, 86, 99, 101, 107, 109, 110
Database, 96
Definitions, 23–25, 71, 86, 96, 111
Descriptions, 9, 24, 30, 78, 96

E

Earth, 10, 12, 29, 42, 43, 45, 64, 75, 79, 81, 84, 85, 97, 99, 100
Entities
   abstract, 46
   bona-fide, 51, 56, 73
   cultural, 72
   fiat, 33, 73

geographical, 7, 10, 12, 16, 21, 28, 31, 32, 34, 40–46, 49, 56, 63–73, 75, 76, 79–87, 93, 95, 99, 101, 107, 108, 123
  historical, 82
  mathematical, 52, 86
  non-existent, 81
  physical, 67
  spatial, 40, 64
Epistemology, 5, 12
Events, 5, 27, 30, 31, 40, 64, 67, 78, 94, 109, 110, 115
Existence, 9, 13, 16, 23, 43–46, 49–55, 60, 65, 67, 71, 72, 81, 82, 87, 113
Experience, 4, 8, 15, 28, 29, 112, 119
Experiments, 11, 12, 29–34, 56
Explanation, 24, 31, 59, 63, 64, 66, 72

F

Facts, 10, 54, 67, 71, 78, 109, 115–118
Field, 10, 13, 22, 55, 63, 79, 80, 86, 87, 95, 98, 124
Folk disciplines, 29

G

Geographical Information Systems (GIS), 7, 8, 28, 52, 59, 70, 74, 86, 87, 95, 123
Geography
  academic, 3, 6, 8, 10, 12, 21, 28, 39, 110
  classical, 39, 41, 42, 46, 70, 85, 86, 103
  common-sense (CSG), 29, 31, 110, 112, 114
  digital, 8
  empirical, 3, 21
  high, 30
  human, 5, 6, 8, 33, 67, 76, 78, 85, 97, 100
  lower, 30, 97, 103
  non-classical, 39, 41, 46, 70, 103
  physical, 5, 33, 67, 76, 84, 97, 98, 101, 103
  professional, 30, 97, 103
  technical, 98, 103
Geometry, 41, 49, 74, 75, 99, 101
Geo-ontologies, 7, 8, 16, 66, 70, 93, 95–104, 107, 109, 110, 112, 114–119, 123, 124

H

Hierarchy, 34, 54, 84, 96, 111, 112
Human beings, 22, 30, 31, 52, 53, 109
Human geo-ontologies, 93, 98, 100, 101, 103

I

Identification, 53, 58, 59, 68, 71, 74, 75, 85, 96
Identity, 9, 26, 41, 43, 44, 46, 64, 67, 70, 80, 83, 94, 97, 109, 113
Individuation, 46, 49, 51, 60, 67, 71
Information, 21, 22, 29, 79, 95, 96, 100, 101, 107, 109, 110, 117
Informativeness, 31, 95, 109, 114–116, 118, 119
Instances, 6, 31, 32, 65–68, 78–80, 84–86, 96, 100, 110–116, 118, 119, 124
Intentionality, 26, 54, 71
Inventory, 9–11, 26, 27, 64, 65, 68, 69, 72, 82, 86, 95, 99, 108, 112
IT/computer science, 9, 10, 21, 23, 95, 103, 107, 108, 114–116, 118, 119, 124

K

Kinds, 3, 6, 7, 9–11, 13, 21, 26–28, 31–34, 40, 41, 43, 46, 49–51, 53–60, 63–67, 70–72, 74, 75, 77–79, 84, 86, 87, 95, 101, 103, 108, 110–119, 123, 124
Knowledge, 4, 5, 8–10, 12, 15, 22–26, 30, 74, 75, 95, 96, 109, 110

L

Language, 9, 15, 22–26, 28, 31, 32, 34, 55, 56, 63, 65, 71–73, 75–77, 82, 86, 87, 96, 99–101, 123
Levels, 26–28, 54, 56, 58, 59, 63, 69, 73, 107, 110, 114, 115, 119, 124
Location, 28, 40–46, 55, 60, 70, 71, 82, 83, 100, 101, 115, 118
Logic, 96
Logical theory, 9, 23, 25, 108

M

Maps, 12, 28, 33, 34, 42, 43, 64, 68–70, 86, 99, 100
Mereology, 39–41, 83
Metaphysics, 12, 13, 26, 27, 51
Methods, 5–7, 10, 12, 14, 25, 26, 28, 31, 86
Models, 5, 6, 10, 12, 23, 42–46, 66, 75, 79, 101, 110, 111
Modes, 8, 109, 113–115

N

Name
  common, 76, 77
  proper, 76, 77

Non-experts, 11, 12, 21, 28–31, 109, 110, 112

O

Objects, 7, 10, 12–14, 16, 24, 25, 27, 29–31, 33, 34, 40, 41, 50, 51, 56, 64, 65, 67, 69–71, 73–75, 78, 79, 83, 99, 109–114
Ontology
  analytical, 10, 12, 26, 50
  continental, 7
  formal, 26, 96, 112
  IT/computer, 3, 9, 10, 21, 23, 25, 95, 111, 112, 119, 123
  material, 26
  of geography, 3, 6–8, 12, 15, 21, 25, 28, 39, 40, 63, 97, 98, 123, 124
  philosophical, 3, 6, 21, 25, 28, 39, 119, 123
  regional, 27, 28

P

Perception, 15, 29, 30
Perspectivism, 65, 119, 123, 124
Phenomenology, 5, 15
Philosophy of geography, 4, 5
Philosophy of language, 26
Physical geo-ontologies, 98, 100, 101, 103
Physics, 25, 28, 30, 69, 119
Places, 4, 82–84, 94, 101
Position, 16, 25, 26, 28, 50, 55, 57, 66, 67, 72, 79, 99, 115–117, 119, 124
Practices, 8, 14, 15, 56, 58, 59, 82
Processes, 7, 10, 16, 24, 27, 30, 31, 78, 87, 94, 97, 109
Properties, 7, 10, 12, 13, 24, 27, 29, 31, 42, 51, 52, 59, 69, 71, 72, 74, 78, 79, 85, 96, 99–101, 103, 109–114, 118
Psychology, 30

R

Realism, 12, 15, 25, 67
Reality, 4–6, 9, 10, 12, 15, 21, 23–31, 34, 39, 52, 53, 56, 65, 69, 71, 77, 80, 82, 86, 87, 108, 123, 124
Relations, 5, 7, 9–14, 23–28, 31, 40, 41, 46, 64, 65, 68, 70, 74–76, 78–80, 82–85, 99, 100, 108–119, 124
Repeated applicability, 95, 99, 100, 109, 119

Representations, 15, 21–23, 28, 39–43, 46, 57, 64, 68, 70, 74, 77, 80, 93, 95, 96, 98, 100, 107, 108, 110, 115, 123

S

Schema, 99, 102
Scientific disciplines, 26–29, 86
Semantics, 10, 21, 22, 25, 74, 75, 77, 95, 99, 100, 107–109
Semantic web, 10, 21, 22, 95, 99, 100, 108, 109
Slots, 24, 111, 112, 114–116, 118, 119, 124
Social sciences, 33, 93
Software, 8, 21, 22, 28, 29, 110
Space, 5, 8, 9, 13, 15, 29–31, 40, 41, 54, 55, 64, 69–71, 80, 81, 85, 86, 93–95, 110
Spatial geo-ontologies, 98, 99
Spatiality, 8, 9, 93–95
Spatial regions, 39, 41–45, 70
Spatial representation, 21, 28, 39–41, 46, 57, 93, 98, 123
Specifications, 9, 10, 21, 23, 25, 41, 66, 71, 80, 108
Structure, 9, 10, 23–27, 30, 31, 40, 54, 56, 63, 71, 73, 78, 84, 86, 96, 108, 111–113, 115, 116
Subjects, 8, 9, 11, 12, 14, 15, 21, 24, 29, 31–34, 46, 56, 68, 87, 93, 94, 96, 98, 124
Systems, 22, 29, 31, 53, 69, 79, 94–96, 98, 100, 110, 114, 116

T

Taxonomies, 25, 49, 51, 54, 55, 58, 71, 72, 74, 93, 95, 96, 98–101, 103
Territory, 45, 46, 50, 68, 81–83, 97, 100, 101
Theories, 4–6, 8, 10, 12, 13, 15, 26, 27, 29, 30, 40, 44, 53, 56, 72, 80, 81, 83, 97, 108, 110
Theory of spatial location, 40, 41, 70
Time, 9, 13, 14, 16, 29–31, 41, 43, 44, 53–56, 63, 64, 67, 71–73, 79–84, 87, 100, 114, 118, 123
Tools, 7, 8, 11–14, 16, 26, 39–41, 63, 80, 86, 87, 123
Topology, 39–41, 81, 83, 101
Turn
  digital, 8
  ontological, 25
  spatial, 86, 93

U
Users, 22, 24, 28, 30, 94, 109, 110, 112

V
Vagueness, 77
Variantism, 115, 116, 123, 124

Vocabulary, 23–25, 96, 99, 101

W
World
  geographic, 12, 70, 71
  mesoscopic, 28, 39, 108

Printed in the United States
by Baker & Taylor Publisher Services